HYPERSPECTRAL IMAGING IN AGRICULTURE, FOOD AND ENVIRONMENT

Edited by **Alejandro Isabel Luna Maldonado, Humberto Rodríguez Fuentes** and **Juan Antonio Vidales Contreras**

Hyperspectral Imaging in Agriculture, Food and Environment
http://dx.doi.org/10.5772/intechopen.70213
Edited by Alejandro Isabel Luna Maldonado, Humberto Rodríguez Fuentes and Juan Antonio Vidales Contreras

Contributors

Sergey Stankevich, Mykola Kharytonov, Anna Kozlova, Vadym Korovin, Mykhailo Svidenyuk, Alexander Valyaev, Jie Chen, Min Zhao, Yankun Peng, Wenxiu Wang, Daniel Farkas, Fartash Vasefi, Nicholas Booth, Hesam Hafizi, Humberto Rodríguez-Fuentes, Alejandro Isabel Luna Maldonado, Juan Antonio Vidales Contreras, Baohua Zhang, Xiaona Li, Ruolan Li, Mengyu Wang, Yaru Liu, Jun Zhou, Zeng Li, Susanto Rahardja, Tatiana Anatolievna Sheremeteva,, Andrey Malov, Gennadiy Filippov, Uziel Francisco Grajeda González, Humberto Rodriguez-Fuentes, Ernesto Alonso Contreras Salazar, Hector Flores Breceda

First published in London, United Kingdom, 2018 by IntechOpen
IntechOpen is the global imprint of INTECHOPEN LIMITED, registered in England and Wales, registration number: 11086078, The Shard, 25th floor, 32 London Bridge Street
London, SE19SG – United Kingdom
Printed in Croatia

British Library Cataloguing-in-Publication Data
A catalogue record for this book is available from the British Library

Additional hard copies can be obtained from orders@intechopen.com

Hyperspectral Imaging in Agriculture, Food and Environment, Edited by Alejandro Isabel Luna Maldonado, Humberto Rodríguez Fuentes and Juan Antonio Vidales Contreras
p. cm.
Print ISBN 978-1-78923-290-5
Online ISBN 978-1-78923-291-2

For EU product safety concerns:
IN TECH d.o.o., Prolaz Marije Krucifikse Kozulić 3, 51000 Rijeka, Croatia,
info@intechopen.com or visit our website at intechopen.com.

Meet the editors

Dr. Alejandro Isabel Luna Maldonado received his PhD degree in Agricultural Science from Kyushu University in 2009. He began his career as a lecturer of Agricultural Engineering at the Autonomous University of Nuevo Leon in 1992. Prof. Luna Maldonado became an assistant professor in 1996, an associate professor in 2014, and a professor in 2018. He is the department head of the Agricultural and Food Engineering since 2017. He is a member of the Mexican National Council for Science and Technology since 2010, Japanese Society of Agricultural Machinery since 2007, and American Society for Agricultural and Biological Engineers.

Dr. Humberto Rodríguez-Fuentes received his PhD degree in Soil and Water from the Autonomous University of Nuevo Leon in 1997. He specializes in biosystems engineering, environmental science, heavy metals, hydroponics, plant nutrition, and plant factory. He began his career as an assistant professor of Agronomy and Agricultural Engineering at the Autonomous University of Nuevo Leon in 1980 and held research positions about soil fertility and pollution, hydroponics, and heavy metals in foods in the same university. Prof. Rodriguez-Fuentes became an associate professor in 2010 and a professor in 2017. He is a member of the Mexican National Council for Science and Technology since 1997.

Dr. Juan Antonio Vidales-Contreras obtained his PhD degree in Soil, Water, and Environmental Science from the University of Arizona in 2001. He specializes in biosystems engineering, environmental bioremediation, hydrology, and wastewater treatment. He began his career as an assistant professor of agricultural engineering at the Autonomous University of Nuevo Leon in 1985 and held research positions about hydraulics, hydrology, soil and water bioremediation, and food physical chemistry in the same university. Prof. Vidales-Contreras became an associate professor in 2009 and a professor since 2014. He is a member of the Mexican National Council for Science and Technology since 2009 and International Water Association since 2003.

Contents

Preface IX

Section 1 Introduction 1

Chapter 1 **Introductory Chapter: Trends on Hyperspectral Imaging Development 3**
Alejandro Isabel Luna Maldonado, Humberto Rodríguez Fuentes and Juan Antonio Vidales Contreras

Section 2 Hyperspectral Imaging in Food 9

Chapter 2 **Multimode Hyperspectral Imaging for Food Quality and Safety 11**
Fartash Vasefi, Nicholas Booth, Hesam Hafizi and Daniel L. Farkas

Chapter 3 **Hyperspectral Imaging and Their Applications in the Nondestructive Quality Assessment of Fruits and Vegetables 27**
Xiaona Li, Ruolan Li, Mengyu Wang, Yaru Liu, Baohua Zhang and Jun Zhou

Chapter 4 **Hyperspectral Imaging for Assessing Quality and Safety of Meat 65**
Wenxiu Wang and Yankun Peng

Chapter 5 **Models Fitting to Pattern Recognition in Hyperspectral Images 85**
Uziel Francisco Grajeda-González, Alejandro Isabel Luna-Maldonado, Humberto Rodriguez-Fuentes, Juan Antonio Vidales-Contreras, Ernesto Alonso Contreras-Salazar and Héctor Flores-Breceda

Section 3 Hyperspectral Imaging in Agriculture 99

Chapter 6 **Sequential Classification of Hyperspectral Images 101**
Min Zhao and Jie Chen

Section 4 Hyperspectral Imaging in Environment 119

Chapter 7 **Soil Contamination Mapping with Hyperspectral Imagery: Pre-Dnieper Chemical Plant (Ukraine) Case Study 121**
Sergey A. Stankevich, Mykola M. Kharytonov, Anna A. Kozlova, Vadym Yu. Korovin, Mykhailo O. Svidenyuk and Alexander M. Valyaev

Section 5 Hyperspectral Imaging Methods 137

Chapter 8 **Using the Target-Visualization Method to Process Hyperspectral Images 139**
Tatiana A. Sheremeteva, Gennadiy N. Filippov and Andrey M. Malov

Chapter 9 **Graph Construction for Hyperspectral Data Unmixing 159**
Zeng Li, Jie Chen and Susanto Rahardja

Preface

Hyperspectral imaging (HSI) is the set of activities by which images are captured and spectral radiance values are assigned to each pixel, through the range of wavelengths of the electromagnetic spectrum under visible and infrared regions. HSI sensors provide information on hundreds of narrow wavelength bands of composite. The pixels are sorted and characterized using statistical analysis and software to classify between groups of pixels. The data from each wavelength band is connected to a three-dimensional hyperspectral imaging (data hypercube) for processing and analysis. Hypercube is composed of layers and each layer represents data at a specific wavelength. HSI is quite important in food quality and assurance, agricultural practices, and environmental quality.

Hopefully, this book will serve as a handbook for students, researchers, and practitioners in HSI and inspire some future research ideas by identifying potential research directions. The book consists of nine chapters, each focusing on a certain aspect of the problem. Within every chapter, the reader will be given an overview of the background information on the subject at hand and in many cases a description of the authors' original proposed solution. The chapters in this book are sorted alphabetically, according to the first author's surname. They should give the reader a general idea where the current research efforts are heading, both within HSI area itself and in interdisciplinary approaches.

Chapter 1 introduces trends on HSI development on fruits and vegetables and meat and tea safety, which can be inspected in-line by HSI not only for increasing the quality and safety of food products but also for offering significant earnings to food processors, with low cost and fast detection of microorganisms. The authors also present HSI techniques for agricultural practices, which are very important in crop monitoring for soil nutrients, plant water stress, disease, insect attack, estimation of crop yield, and pollution tracking.

Chapter 2 introduces multimode hyperspectral imaging for food quality and safety because there is a great need to assess the composition of food, quantitatively and reproducibly, to avoid any unintended scenarios, ranging from a product not being quite what it is stated to be intentional adulteration to random contamination.

Chapter 3 considers hyperspectral imaging and their applications in the nondestructive quality assessment of fruits and vegetables. The authors report HSI as a detailed overview of the introduction, latest developments, and applications of hyperspectral imaging in the nondestructive assessment of fruits and vegetables. Additionally, they include principal components, basic theories, and corresponding processing and analytical methods.

Chapter 4 reports HSI for assessing quality and safety of meat. The authors present quality attributes including meat sensory attributes, chemical attributes, and technological attributes. In addition, they show the advantages, disadvantages, and problems to be tackled facing the HSI.

Chapter 5 considers model fitting to pattern recognition in hyperspectral images. The authors propose a new algorithm for *Salmonella typhimurium* detection on tomato surfaces in a visible range (400–1000 nm). Gaussian model was used to take out a model that could calculate its definite integral, and the result of this algorithm is the area under curve (AUC), which gives a quantitative approach of spectral signatures.

In Chapter 6, a sequential classification of hyperspectral images is analyzed. The authors consider the hyperspectral classification problems with the consideration of sequential data collection. The related techniques include data normalization, dimension reduction, classification, spatial information integration, etc., and the way to accommodate these techniques into the context of sequential data collecting and processing.

Chapter 7 presents soil contamination mapping with HSI on Prydniprovskiy Chemical Plant in Ukraine, which is an issue of severe importance for Ukraine. The authors introduce regular radioecological observations and up-to-date contamination mapping based on advanced HSI geoinformational techniques and showcase an ability to prepare for, respond to, and manage potential adverse effects from pollution with radionuclides and heavy metals.

Chapter 8 reports the target visualization method to process a hyperspectral image. The authors introduce how visualization of data sets may be useful, for object recognition, by using additional nonformalized external attributes. They convert a hyperspectral image into a single-channel synthesized image in gray scale, on which the objects of interest for the problem under consideration are selected.

Chapter 9 introduces a novel graph construction for hyperspectral data unmixing. The authors present graph construction for hyperspectral data and associated unmixing methods based on graph regularization. The authors review fundamental elements of graph and present different ways to construct graphs in both spatial and spectral senses for hyperspectral images.

Prof. Alejandro Isabel Luna Maldonado
Agricultural and Food Engineering at the Autonomous University of Nuevo León
Escobedo, Nuevo León, Mexico

Dr. Humberto Rodríguez Fuentes
Agricultural and Food Engineering at the Autonomous University of Nuevo León
Escobedo, Nuevo León, Mexico

Dr. Juan Antonio Vidales Contreras
Agricultural and Food Engineering at the Autonomous University of Nuevo León
Escobedo, Nuevo León, Mexico

Section 1

Introduction

Chapter 1

Introductory Chapter: Trends on Hyperspectral Imaging Development

Alejandro Isabel Luna Maldonado,
Humberto Rodríguez Fuentes and
Juan Antonio Vidales Contreras

Additional information is available at the end of the chapter

http://dx.doi.org/10.5772/intechopen.76998

1. Introduction

Hyperspectral imaging (HSI) is the set of activities by which images are captured and spectral radiance values assigned to each pixel through the range of wavelengths of the electromagnetic spectrum under visible and infrared regions. HSI sensors provide information on hundreds of narrow wavelength bands of composite. The pixels are sorted and characterized using statistical analysis and software to classify among groups of pixels. The data from each wavelength band are connecting into a three-dimensional hyperspectral (data hypercube) for processing and analysis. Hypercube is composed of layers, and each layer represents data at a specific wavelength. HSI is quite important in food quality and assurance, agricultural practices and environmental quality.

2. HSI in assurance of food quality

Nowadays, consumers are interested in food quality and safety assurance. Fruits and vegetables can be inspected in-line by HSI for increasing the quality and safety of food products but also to offer significant earnings to food processors [1]. Spectral variations due to morphological changes of most fruits and vegetables decrease the prediction of models; in addition, the interferences that specimens possess might affect the classification accuracy, therefore imaging techniques in dealing with morphological effects are needed. Another demand is the automatic recognition of representative region of interest based on computer software to improve model efficiency. HSI systems with low-cost and fast-detecting properties are provided [2]. An effective pixel-based apple bruise region extraction method has been proposed to obtain the complete bruise region. The hyperspectral images of 60 apples were obtained at 0, 12, and 18 h during an experiment. Principal component analysis (PCA) eliminates repetitious data of hypercubes. Random Forest (RF) model obtained high and steady classification accuracy. The mean accuracy of bruise extraction models reached 99.9 [3]. Color parameters (L*, a* and b*), firmness, and soluble solid content (SSC) have been quantified by HSI in the visible and near infrared (VNIR) regions between 600 and 975 nm and the short wave near infrared (SWIR) region between 865 and 1610 nm. SSC can be exactly predicted by SWIR hyperspectral imaging with than 0.8, while L* and a* adjusted better with VNIR hyperspectral imaging displayed correlation coefficients greater than 0.7 for [4]. Near infrared (NIR) hyperspectral imaging can classify among maize kernels of varying hardness and between fungal infected and sound kernels [5].

For the quick and nondestructive detection of microbial decay in muscle of beef, pork, poultry, fish, and so on, techniques have been used such as visible and near-infrared spectroscopy, Fourier transform infrared spectroscopy, fluorescence spectroscopy, Raman spectroscopy, and hyperspectral imaging. When those techniques are combined with chemometric analysis, spectral preprocessing and modeling methods are successfully developed for the determination of total viable count, aerobic plate count, *Enterobacteriaceae, Pseudomonas, Escherichia coli,* and lactic acid bacteria loads in muscle [6].

Starch content in adulterated fresh cheese has been measured using HSI. In a research, adulterated fresh cheese was prepared using concentrations of starch of 0.055–12.705 mg g^{-1}; afterwards, HSI images in the range of 200–1000 nm, distributed in 101 bands were acquired. A partial least square regression (PLSR) model of starch content was obtained with a determination coefficient (R^2) of 0.9915 and a root mean square error of cross-validation (RMSECV) of 0.3979. With five variables, a correlation coefficient of validation (r) of 0.8321 and a mean square error prediction (RMSEP) of 1.3515 was found for a reduced model [7].

Haugh unit (HU) index is a measure of the quality of the albumen in various studies on egg quality. HU is a destructive test of specimen and correctly reflects the batch of eggs being processed. In a study, fresh eggs were stored at 25°C and measured after storage for 0, 4, 7, 10, 14, 18 and 21 days by HSI system in the wavelength range of 900–1700 nm and compared to HU for each egg. A calibration model for HU initially used PLSR and then cross-validation was performed and a coefficient of determination (R^2) of 0.91 and root mean square error of calibration (RMSEC) of 4.58 was obtained; however, displayed colors of acquired image of eggs were different correspond to the freshness of the eggs based on HU [8].

Dairy product companies are demanding systems for quantifying and qualifying differences between milk powders. Hyperspectral imaging (HSI) has been used to distinguish between milk powders manufactured in factories and of differing practical and useful qualities, for instance, dispersibility. HSI and multivariate analysis techniques such as principal component analysis (PCA) and partial least squares (PLS) regression were performed. The PCA results exhibited differences in the first and second principal components. The PLS technique showed that HSI information could be used to forecast the dispersibility parameter and then establish significant correlations between hyperspectral images and crucial quality attributes of milk powder either on, or at line in close to real time [9].

Herbal tea demand is increasing because of consumers coming to know of its health benefits. Chromatographical techniques require destructive sample preparation using solvents; therefore, HSI could be a nondestructive alternative method. In a research, HSI pushbroom system captured images of the raw material and tea blends by SisuChema SWIR (short wave infrared). Subsequently, the images were analyzed using multivariate analysis software. PCA revealed 54.2% chemical variation between *S. tortuosum* and *C. genistoides* raw materials. A partial least squares-discriminant analysis (PLS-DA) model had confidence prediction of 95.8% and it was possible to visualize the tea blend constituents (based on pixel classification) as *S. tortuosum* and *C. genistoides* and quantitatively predicted *C. genistoides* as the major constituent (>97%) while *S. tortuosum* was existent in lower amounts (<3%) [10].

Advanced preprocessing methods for denoising that possess high efficiency and high exactitude are appearing to improve the predicting accuracy for using hyperspectral images in food quality evaluation and analysis. Adaptive filters have been developed for applications since they can steadily adjust itself to the changing imaging environment [11]. Hyperspectral pansharpening method has been used with high frequency layer of each band of the hyperspectral image as the guidance image of the guided filter for extraction of spatial details from both the panchromatic image and the hyperspectral image. The total spatial attributes are added in the end into each band of the HS image low frequency layer to generate the last image [12].

On the other hand, hyperspectral fluorescence imaging (HSFI) method has been used to evaluate quality and safety of food since it combines the advantages of both hyperspectral imaging and fluorescence spectroscopy. However, it cannot be said that HSFI is very effective for measuring quality attributes. The potential of this technology for food and agricultural product quality and safety in online inspection will improve rapidly with advances in optical sensing and computer systems [13].

3. HSI in assurance of agricultural practices

The development of hyperspectral imaging systems, both aerial and ground, has been very important in crop monitoring for nutrients, water stress, disease, insect attack and estimation of crop yield in smart agriculture.

The remote perception of water stress in a citrus orchard have been researched using leaf measurements of chlorophyll fluorescence and Photochemical Reflectance Index (PRI) data, seasonal time-series of crown temperature and PRI, and high-resolution airborne imagery [14]. The miniaturization technology has supplied markets with hyperspectral imagers operating on frame format, which is highly attractive for unmanned aerial vehicle (UAV) based remote sensing, because it provides better stability and the likelihood to grab stereoscopic data sets, bringing in a possibility for three dimensional hyperspectral object reconstruction [15]. In some studies, UAV has been performed to acquire RGB images for vegetation analysis [16]. Liquid crystal tunable filter (LCTF)-based hyperspectral imaging system transmitted selected wavelengths without the requirement to exchange optical filters from UAV and measured 14 different ground objects in vegetative areas. Additionally, the machine learning (ML) approach using a support vector machine (SVM) model reached a classification accuracy of 94.5% in vegetated areas [17].

In recent research, automated remote sensing procedures have been developed, assessed, and compared based on novel, low-cost HSI system for the identification of beetle infestations in barks at the individual tree level in urban forests achieving an overall accuracy of 81% (kappa: 0.70), compared to the aircraft results of 73% (kappa: 0.56) in a smaller sub-area [18].

4. HSI in assurance of environmental quality

HSI is rapidly becoming a key tool for pollution tracking changes in the environment. Hyperspectral microscopy (HM) has been explored for nanotoxicity studies of materials in a more native state and truer to conditions of biomedical pertinence. Additionally, HM had potential and found its earliest macroscopic applications in geologic surveying. However, analysis of air or water samples is constrained by the challenge of immobilizing particles [19]. A study based on Hyperspectral Imaging (HSI), was developed to establish an efficient method to characterize marine microplastic litter. Reliable information on abundance, size, shape and polymer type for the whole ensemble of plastic particles in each sample was retrieved from hyperspectral images [20]. A new algorithm has been evaluated using the Hyperspectral Imager for Coastal Ocean (HICO). The hyperspectral vicarious calibration was applied to HICO, showing the validity and consistency of HICO's ocean color products [21]. HSI can arrange pixel providing a lot of potential for material characterization. A study demonstrated that HSI is possible for recognition of pigments [22]. Modern studies of heavy metal pollution of soils have been focused on the hyperspectral reflectance of typical metals in soils and in plants measured either *in situ* or in the laboratory. Most of these studies used wavebands lie within the visible near infrared range of the spectrum, especially the red edge. Metals detection must rely on their co-variation with the spectrally responsive metals or organic matter in the soils [23].

Acknowledgements

As the topic editors, we would like to thank all our colleagues who contributed their chapters. We are especially grateful for excellent technical support provided by the publishing process managers. The editors express their sincere gratitude to PAICYT UANL and National Council of Science and Technology for the support. The authors also wish to give sincere thanks to Ph.D. Alejandro Sergio Del Bosque González for their comments on the manuscript and support during this research and publication.

Author details

Alejandro Isabel Luna Maldonado*, Humberto Rodríguez Fuentes* and Juan Antonio Vidales Contreras*

*Address all correspondence to: alejandro.lunaml@uanl.edu.mx, juan.vidalescn@uanl.edu.mx and humberto.rodriguezfn@uanl.edu.mx

Faculty of Agriculture, Food and Agricultural Engineering, Autonomous University of Nuevo Leon, General Escobedo, Mexico

References

[1] Gallagher MS, Mahajan PV. The stability and shelf life of fruit and vegetables. In: Food and Beverage Stability and Shelf Life. Woodhead Publishing Ltd.; 2011. pp. 641-656

[2] Pu YY, Feng YZ, Sun DW. Recent progress of hyperspectral imaging on quality and safety inspection of fruits and vegetables: A review. Comprehensive Reviews in Food Science and Food Safety. 2015;**14**(2):176-188

[3] Che W, Sun L, Zhang Q, Tan W, Ye D, Zhang D, Liu Y. Pixel based bruise region extraction of apple using Vis-NIR hyperspectral imaging. Computers and Electronics in Agriculture. 2018;**146**:12-21

[4] Li B, Cobo-Medina M, Lecourt J, Harrison NB, Harrison RJ, Cross JV. Application of hyperspectral imaging for nondestructive measurement of plum quality attributes. Postharvest Biology and Technology. 2018;**141**:8-15

[5] Williams PJ. Near infrared (NIR) hyperspectral imaging for evaluation of whole maize kernels: Chemometrics for exploration and classification [Doctoral dissertation]. Stellenbosch: University of Stellenbosch; 2009

[6] Cheng JH, Sun DW. Recent applications of spectroscopic and hyperspectral imaging techniques with chemometric analysis for rapid inspection of microbial spoilage in muscle foods. Comprehensive Reviews in Food Science and Food Safety. 2015;**14**(4):478-490

[7] Barreto A, Cruz-Tirado JP, Siche R, Quevedo R. Determination of starch content in adulterated fresh cheese using hyperspectral imaging. Food Bioscience. 2018;**21**:14-19

[8] Suktanarak S, Teerachaichayut S. Non-destructive quality assessment of hens' eggs using hyperspectral images. Journal of Food Engineering. 2017;**215**:97-103

[9] Munir MT, Wilson DI, Yu W, Young BR. An evaluation of hyperspectral imaging for characterising milk powders. Journal of Food Engineering. 2018;**221**:1-10

[10] Sandasi M, Chen W, Vermaak I, Viljoen A. Non-destructive quality assessment of herbal tea blends using hyperspectral imaging. Phytochemistry Letters. 2018;**24**:94-101

[11] Dai Q, Sun DW, Cheng JH, Pu H, Zeng XA, Xiong Z. Recent advances in de-noising methods and their applications in hyperspectral image processing for the food industry. Comprehensive Reviews in Food Science and Food Safety. 2014;**13**(6):1207-1218

[12] Dong W, Xiao S, Li Y. Hyperspectral pansharpening based on guided filter and Gaussian filter. Journal of Visual Communication Image Representation. 2018;**52**:151-158

[13] Zhang R, Ying Y, Rao X, Li J. Quality and safety assessment of food and agricultural products by hyperspectral fluorescence imaging. Journal of the Science of Food and Agriculture. 2012;**92**(12):2397-2408

[14] Honkavaara E, Rosnell T, Oliveira R, Tommaselli A. Band registration of tuneable frame format hyperspectral UAV imagers in complex scenes. ISPRS Journal of Photogrammetry and Remote Sensing. 2017;**134**:96-109

[15] Zarco-Tejada PJ, González-Dugo V, Berni JA. Fluorescence, temperature and narrow-band indices acquired from a UAV platform for water stress detection using a micro-hyperspectral imager and a thermal camera. Remote Sensing of Environment. 2012; **117**:322-337

[16] Senthilnath J, Kandukuri M, Dokania A, Ramesh KN. Application of UAV imaging platform for vegetation analysis based on spectral-spatial methods. Computers and Electronics in Agriculture. 2017;**140**:8-24

[17] Ishida T, Kurihara J, Viray FA, Namuco SB, Paringit EC, Perez GJ, Takahashi Y, Marciano JJ. A novel approach for vegetation classification using UAV-based hyperspectral imaging. Computers and Electronics in Agriculture. 2018;**144**:80-85

[18] Näsi R, Honkavaara E, Blomqvist M, Lyytikäinen-Saarenmaa P, Hakala T, Viljanen N, Kantola T, Holopainen M. Remote sensing of bark beetle damage in urban forests at individual tree level using a novel hyperspectral camera from UAV and aircraft. Urban Forestry & Urban Greening. 2018;**30**:72-83

[19] Roth GA, Tahiliani S, Neu-Baker NM, Brenner SA. Hyperspectral microscopy as an analytical tool for nanomaterials. Wiley Interdisciplinary Reviews: Nanomedicine and Nanobiotechnology. 2015;**7**(4):565-579

[20] Serranti S, Palmieri R, Bonifazi G, Cózar A. Characterization of microplastic litter from oceans by an innovative approach based on hyperspectral imaging. Waste Management. 2018. In Press

[21] Ibrahim A, Franz B, Ahmad Z, Healy R, Knobelspiesse K, Gao BC, Proctror C, Zhai PW. Atmospheric correction for hyperspectral ocean color retrieval with application to the hyperspectral imager for the coastal ocean (HICO). Remote Sensing of Environment. 2018;**204**:60-75

[22] Grabowski B, Masarczyk W, Głomb P, Mendys A. Automatic pigment identification from hyperspectral data. Journal of Cultural Heritage. 2018;**31**:1-12

[23] Wang F, Gao J, Zha Y. Hyperspectral sensing of heavy metals in soil and vegetation: Feasibility and challenges. ISPRS Journal of Photogrammetry and Remote Sensing. 2018; **136**:73-84

Section 2

Hyperspectral Imaging in Food

Chapter 2

Multimode Hyperspectral Imaging for Food Quality and Safety

Fartash Vasefi, Nicholas Booth, Hesam Hafizi and Daniel L. Farkas

Additional information is available at the end of the chapter

http://dx.doi.org/10.5772/intechopen.76358

Abstract

Food safety and quality are becoming progressively important, and a failure to implement monitoring processes and identify anomalies in composition, production, and distribution can lead to severe financial and customer health damages. If consumers were uncertain about food safety and quality, the impact could be profound; hence, we need better ways of minimizing such risks. On the data management side, the rise of artificial intelligence, data analytics, the Internet of Things, and blockchain all provide enormous opportunities for supply chain management and liability management, but the impact of any approach starts with the quality of the relevant data. Here, we present state-of-the-art spectroscopic technologies including hyperspectral reflectance, fluorescence imaging as well as Raman spectroscopy, and speckle imaging that are all validated for food safety and quality applications. We believe a multimode approach comprising of a number of these synergetic optical detection modes is needed for the highest performance. We present a plan where our implementations reflect this concept through a multimode tabletop system in the sense that a large, real-time production-level device would be based on more modes than this mid-level one, while a handheld, portable unit may only address fewer challenges, but with a lower cost and size.

Keywords: multimode optical imaging, food contamination, hyperspectral imaging, food quality, multimode data management, machine learning

1. Introduction

There is a great need to assess the composition of food, quantitatively and reproducibly, in order to avoid any unintended scenarios, ranging from a product not being quite what it is stated to be (e.g., lesser quality fish or olive oil) to intentional adulteration (including by terrorist intent) to random contamination (such as by bacteria that can be lethal). These constitute the application domain of, respectively, food quality, food defense, and food safety. Given the place food occupies in society, and the possible extreme implications of any negative events, there is great interest in bringing the best testing to the task of ensuring the quality and safety of our food supply. Unfortunately, some of the currently prevalent methods (molecular/biochemical/biophysical, such as polymerase chain reaction (PCR), chromatography, mass spectrometry, etc.) are intrinsically too slow to yield results in real time, and also rely on random and very sparse sampling. We believe that the power of light as an investigational tool can be brought to the resulting challenge and focus on this possibility here.

Optical imaging is an approach rapidly growing in popularity and applications due to technological advances that have enabled the production of smaller, less expensive, more efficient, and faster light sources and detectors. These new technologies have facilitated the acquisition of more accurate optical image sets, yielding molecular, structural, and physiological information from targeted samples. There are many different optical measurement techniques used by industry and academic researchers alike, with each technology usually focusing on a specific property of light (intensity, polarization, wavelength, coherence, temporal change, etc.). We believe, however, that no single method can provide the comprehensive analysis of food that is required.

When applied to food samples, the accuracy of optical detection techniques can be limited due to factors such as low penetration depth and lack of contrast, especially for low biomarker concentrations. However, using a strategic combination of multiple optical detection technologies in an optical system that thus becomes *multimode*, the chemical and/or biological detection accuracy can be substantially improved. Each individual detection method can provide a specific and complementary (sometimes even synergetic) piece of information regarding the sample being examined. Thus, by combining a number of these methods, the impact of the individual limitations can be minimized, and their combined strengths may be harnessed to deliver highly specific results.

The advantages of multimode optical imaging include greatly reducing the time required for the initial detection and enumeration of contaminants, with minimal sample preparation, nondestructive evaluation, fast acquisition times, and visualization of the spatial distribution of numerous components simultaneously. These advantages are highly useful in detecting contaminants in food for assessing safety and quality, and the use of multiple modes of detection, properly combined, is essential for effectiveness and performance.

We summarize here optical technologies which are useful in food safety and quality applications, highlighting both successes and limitations, thus underscoring the usefulness of the new, multimode approach we propose.

2. Hyperspectral imaging

Hyperspectral imaging (HSI) is a growing platform technology that functions by integrating conventional imaging and spectroscopy to gain spatial and spectral information from an object [1]. It is capable of capturing reflectance, transmittance, and fluorescence images in the visible and infrared regions with submillimeter spatial resolution [2] and high spectral resolution (10 nm). While HSI was originally developed for remote sensing [3], it has gained popularity in the field of food safety and analysis with new applications reported in fruits and vegetables [4–20, 34, 37, 42], poultry [21–25], and meat [26–28]. Some advantages HSI has in comparison with other techniques such as RGB imaging, NIR spectroscopy, and multicolor imaging include being able to produce spatial and spectral information, multiconstituent information, and sensitivity to minor components [1].

HSI in the near infrared (NIR) can provide chemical composition of red meat such as prediction of fat, protein, and water content of lamb meat [32]. Moreover, this method enables the detection of certain bacteria in food, such as *E. coli* [33]. Fungal growth on food products is of particular concern due to the potential for detrimental effects on population health ranging from allergic reactions and respiratory problems to the production of mycotoxins. HSI has been deployed to identify fungal species such as *Aspergillus flavus, Aspergillus parasiticus, Aspergillus niger,* and *Fusarium* spp. which can produce mycotoxins, which are secondary metabolites that are toxic for humans and animals [36, 37].

A common source of contamination for fresh products and other raw materials used to produce food is fecal contamination; hence it would be highly desirable to develop an automatic inspection system for use in the field and on processing lines. Multispectral detection of fecal contamination on apples using HSI imaging was demonstrated by Kim et al. [45]. A HSI system with a range of 450–851 nm was used to examine reflectance images of experimentally contaminated apples. Fecal contamination sites were evaluated using principal component analysis (PCA) with the goal of identifying two to four wavelengths that could be used in an online multispectral imaging system. As shown in **Figure 1**, their results showed that contamination could be identified using either of three wavelengths in the green, red, and NIR regions.

With the use of HSI in the spectral range of 400–1000 nm, *E. coli* loads in grass carp fish have been measured to evaluate microbial spoilage. In 2015, the researchers demonstrated that reflectance HSI in combination with multivariate analysis had the ability to rapidly and noninvasively quantify and visualize the *E. coli* loads in grass carp fish flesh during the spoilage process [35]. Distribution maps, shown in **Figure 2**, were created to allow for visualization of *E. coli* contamination. These distribution maps were vital in that they provided more detailed information of postmortem spoilage development in grass carp flesh. One of the main advantages that HSI has over conventional spectroscopy methods is its ability to visualize distribution maps of the contamination in a pixel-wise manner. By multiplying the regression coefficients of the multiple linear regression model by the spectrum of each pixel in the image, a prediction map was generated for showing the distribution of *E. coli* within the fish flesh. The different *E. coli* loads were represented by different colors from blue to red. As *E. coli* load increased, the color of the images shifted from blue to red, reflecting the growth of bacteria.

In 2013, Feng et al. [36] presented HSI as a nondestructive tool for direct, quantitative determination of Enterobacteriaceae loads on chicken fillets. The authors developed partial least

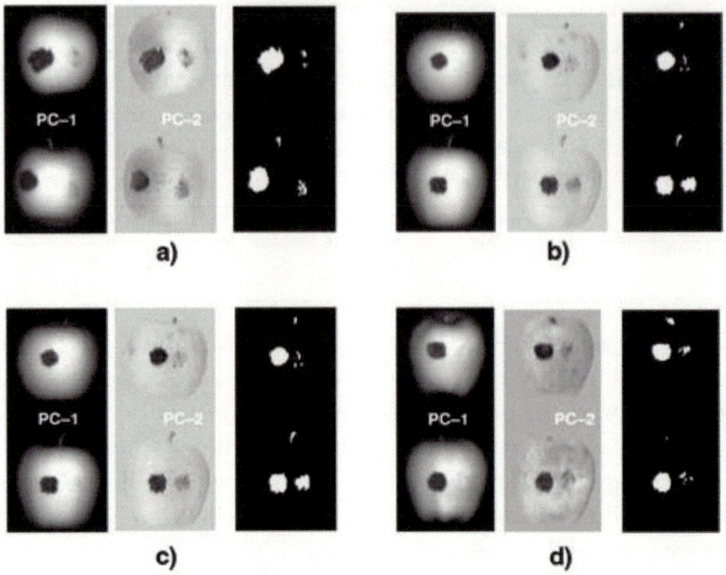

Figure 1. First and second principal component images obtained using 748–851 nm region of the hyperspectral reflectance image data for (A) fuji, (B) gala, (C) golden delicious, and (D) red delicious apples [45].

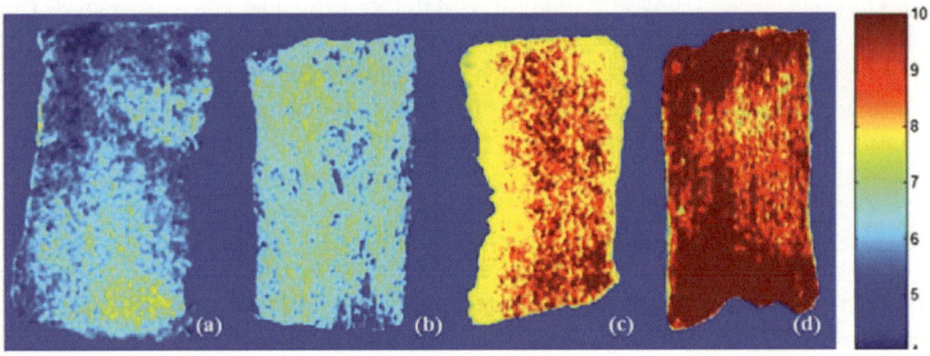

Figure 2. These are examples of distribution maps of *E. coli* loads in fish filets. The distribution maps showed how the level of *E. coli* contamination varied from one sample to the next. A shift in color intensity is seen from blue to red, reflecting the increase in *E. coli* contamination [35].

squares regression (PLSR) models and root mean squared errors. After a simplified model was developed, the PLSR model, it was used for predicting Enterobacteriaceae loads in every pixel of the image acquired from HSI, resulting in a new image called a "prediction map." In this prediction map, a color scale was used to describe the different microbial loads in each spot of the sample. As shown in **Figure 3**, when the microbial loads increase, the images shift from a blue color to a more reddish one, this reflects the growth of bacteria on the chicken fillets.

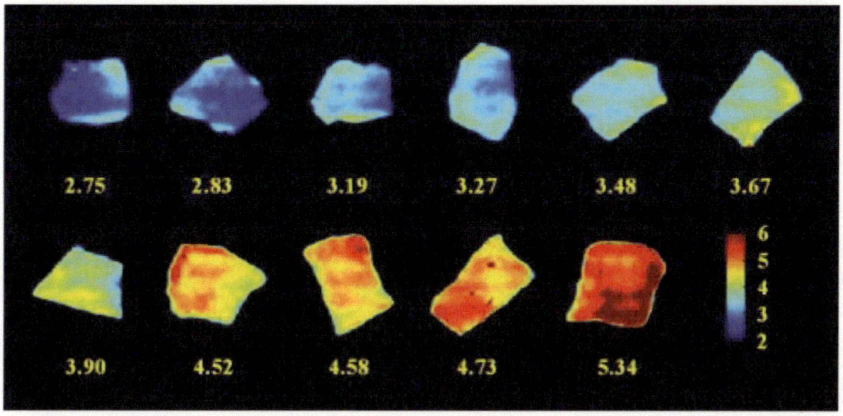

Figure 3. This is an image of a median-filtered prediction map for validation set using the simplified PLSR model built on three wavelengths (930, 1121, and 1345 nm). Values under each sample predict the Enterobacteriaceae counts (in $\log_{10}$ CFU g^{-1}) [36].

Changes in temperature during cold storage of meat products can lead to undesirable microbial growths, which may affect food safety. A study of the spoilage of beef was reported by Peng et al. [41]; in this work, HSI was exploited to measure biochemical changes within the fresh beef. The research demonstrated that HSI showed potential for real-time and nondestructive detection of bacterial spoilage in beef.

Work performed by Barbin et al. [43] used HSI in the near-infrared range (900–1700 nm) to determine the total viable count and psychotropic plate count in chilled pork during storage. NIR hyperspectral images in the reflectance mode were captured every 48 h from each sample. Assuming that meat spoilage is evident at a microbial load of 107 CFU per gram or cm^2, the author's defined a cutoff point of 106 CFU/g as an acceptable threshold of freshness. By examining the spectral information that was obtained from the samples, a difference was observed in the wavelength range between 1300 and 1600 nm, where fresh samples had lower absorbance than spoiled samples (see **Figure 4**). This spectral region is commonly assigned to N-H stretch of proteins (amines and amides) and their interactions with water, and it could suggest the occurrence of proteolytic changes, which are recognized as the main indicator for the onset of spoilage in meat products.

In 2016, Everard et al. [51] presented fluorescence HSI coupled with multivariate image analysis techniques utilized for the detection of fecal contaminates on spinach leaves. Violet fluorescence excitation was provided at 405 nm, and light emission was recorded from 464 to 800 nm. Partial least square discriminant analysis (PLSDA) and wavelength ratio methods were compared for detection accuracy for fecal contamination. The PLSDA model had 19% false positives for nonfresh post storage leaves. A wavelength ratio technique using four wavebands (680, 688, 703, and 723 nm) was successful in identifying 100% of fecal contaminates on both fresh and nonfresh leaves.

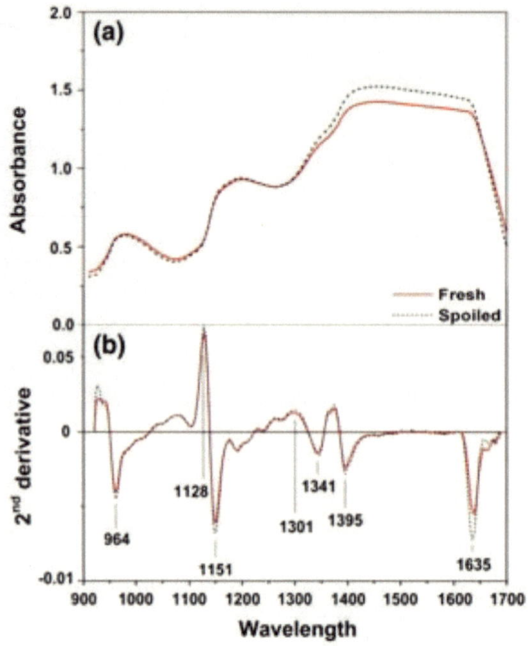

Figure 4. (a) absorbance spectra for fresh and spoiled samples (after 7 days of storage); (b) second derivative spectra for fresh and spoiled samples showing potentially relevant wavelengths [43].

Detection of fecal contamination on cantaloupes using HS fluorescence imagery was demonstrated by Vargas et al. [46]. HS images of cantaloupes artificially contaminated with a range of diluted bovine feces were acquired from 425 to 774 nm in response to ultraviolet-A (320–400 nm) excitation. Evaluation of images at emission peak wavelengths indicated that 675 nm exhibited the greatest contrast between contaminated and untreated surface areas. Two-band ratios compared with the single-band images enhanced the contrast between the fecal contaminated spots and untreated cantaloupe surfaces.

Yang et al. [47] examined methods to classify fecal contamination on leafy greens. They utilized HS fluorescence imaging system with ultraviolet-A excitation (320–400 nm) for detection of bovine fecal contaminants on the abaxial and adaxial surfaces of romaine lettuce and baby spinach leaves. They applied six spots of fecal contamination to each of the 40 lettuce and spinach leaves. Their results showed that for both lettuce and spinach, the detection of fecal matter was best obtained using the ratio of the signal from 666 nm divided by that from 680 nm, (R values of 0.98 for romaine lettuce and 0.96 for baby spinach).

3. Raman spectroscopy and spectral imaging

Raman spectroscopy is a nondestructive spectroscopic technique, based on the vibrational properties of the constituent molecules, that provides molecular information about the sample under examination. The Raman signal results from molecules being excited by a small amount of incident light at a specific wavelength. The remitted light has some of its photons shifted to different wavelengths by the addition or subtraction of vibrational energy from some of the tissue intramolecular bonds [44]. Contrast is achieved when the tissue molecular constituents differ enough that the Raman signals from two tissues have different wavelength distributions. Raman spectral imaging (RSI) intertwines Raman spectroscopy and digital imaging to visualize the composition and structure of a target, thereby having great potential for food safety and analysis [29]. Historically Raman imaging systems have only been able to perform Raman measurement at a microscopic level and were unable to evaluate whole surfaces of individual foods. Recent studies have shown a benchtop point-scanning Raman chemical imaging system designed and developed for food safety research [56]. Raman imaging is a highly specific and sensitive technique as it allows for the detection of particular chemicals at low concentrations, such as detecting melamine particles in dry milk. This technique has wide applications, and due to its specificity, it may help detect contaminants in food products of different sizes.

A study aimed at the detection and differentiation of important food and waterborne bacteria (*E. coli, Staphylococcus epidermidis, Listeria monocytogenes,* and *Enterococcus faecalis*) was performed by Fan et al. [38] using surface-enhanced Raman spectroscopy (SERS) coupled with intracellular nanosilver as SERS substrates. Variations observed in the spectral patterns of bacterial pathogens are due to the different quantity and distribution of cellular components like proteins, phospholipids, nucleic acids, and carbohydrates. SERS coupled with statistical analysis has become very useful in discriminating and detecting bacterial cells, spores, and viruses.

In another study, a portable Raman sensor system was presented with an integrated 671 nm microsystem diode laser as excitation light source for the rapid in situ detection of meat spoilage and bacteria [39]. The system used in this chapter is an example of the reduction in form factor of enabled by recent advances and is comprised of three main components: a handheld measurement head with a laser driver electronics board, the Raman optical bench, and finally, a battery pack. This method was used to rapidly detect meat spoilage in specific pork cuts, *musculus longissimus dorsi* (LD) and *musculus semimembranosus* (SM). The authors were able to determine the total number of mesophilic aerobic microorganisms on the surface of the meat to show possible correlations of the bacterial growth with the measured Raman spectra. In 2007, the food industry faced substantial economic losses following the discovery of melamine, a nitrogen rich chemical, in human and pet foods [48]. In one SERS study which employed SERS-active substrates, the concentration of melamine was measured in wheat gluten, chicken feed, and processed foods such as cake and noodles [49, 50].

4. Speckle imaging

Spoilage and poisoning of food products by microorganisms is a major issue in food safety and human health. As these microorganisms grow and become more active, they cause deterioration of food quality and cause food intoxication. Some of the microorganisms capable of such damage are bacteria, yeast, and mold. As detailed earlier, there have been many different technologies developed to detect harmful microorganisms in food products such as hyperspectral imaging, Raman spectroscopy, and high-performance liquid chromatography. All these methods have certain intrinsic short comings. Factors such as the need for a well-equipped laboratory, high-cost equipment, complicated procedures for sample preparation and long analysis times, and trained professional operators limit their widespread application in the food processing, transportation, marketing, and preservation in various food industries.

A technology that is finding increasing favor by circumventing many of these limitations is laser speckle imaging. Laser speckle imaging has been introduced in this field of application to monitor moving particles in optically inhomogeneous media by analyzing time-varying laser speckle patterns for applications such as measuring meat quality and detecting contaminants. Unlike multiple light scattering in meat which exhibits static and deterministic speckle intensity patterns, light paths associated with the movements of living microorganisms result in time-varying changes in the speckle intensity patterns. Therefore, by detecting the decorrelation in the laser speckle intensity patterns from tissues, the living activities of microorganisms can be detected.

Another advantage of this method is the ability to examine meats sealed with transparent packaging because this method detects time-varying signals in reflected laser beams and transparent plastic does not affect these. Furthermore, the technique can provide rapid assessment as bacterial colonies can be detected within a few seconds [30]. Thus, this method provides an efficient and effective way to detect live bacteria in food products to avoid food toxicity. Speckle imaging systems have been demonstrated to indicate the presence of bacterial colonies and other contaminants in both food and water [31]. Technology such as this may be very effective in the marketplace as food producers or consumers themselves may be able to use them to assess food safety. As mentioned, there are currently several approaches available for detecting low levels of microorganisms in food; however, they require complex equipment, high costs, invasive procedures, and skilled technicians to operate which all act to restrict its widespread adoption and use in the food industry [31].

Work performed by Yoong et al. [53] aimed to detect and quantify various levels of contamination using chicken breast meat samples. The meats contaminated with bacteria had significant decreases in the autocorrelation values over the time lag, whereas the control group (meat treated with a PBS solution) did not show any major changes. The meat treated with a high concentration of bacteria had more significant changes over the time lag compared with the meat treated with a low concentration of bacteria. Moreover, the decrease in the autocorrelation value was proportional to the concentration of the treated bacteria. The measured autocorrelation values were all statistically different from one another ($p < 0.001$), and the decreases in the autocorrelation were proportional to the concentration of bacteria. Thus, the authors were able to show that through various experimental validations, spontaneous bacterial activity caused strong decorrelation in laser speckle dynamics (**Figure 5**).

In 2014, Kim et al. [55] presented a label-free bacterial colony phenotyping technology called bacterial rapid detection using optical scattering technology (BARDOT), which can provide classification for several different types of bacteria. Recent experiments with colonies of *Bacillus* species using speckle imaging show a certain speckle formation that allows for the detection and

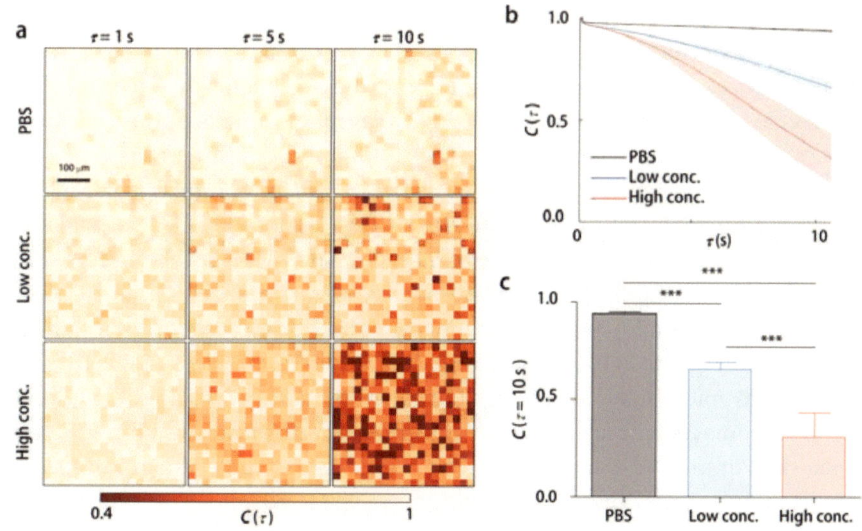

Figure 5. This image illustrates the groups attempt at assessing bacterial activity in meat. (A) shows representative autocorrelation amps in meat treated with various concentrations of bacteria at various time lags. (B) Averaged C(tau) values over the areas in (A) as a function of the time lag. (C) quantification of the autocorrelation values at tau = 10 s [53].

identification of these bacterial species. As the center diameter of the *Bacillus* spp. colony grew from 500 to 900 microns, the average speckle area decreased twofold and the number of small speckles increased sevenfold. As *Bacillus* colonies grow, the average speckle size in the scatter pattern decreases and the number of smaller speckle increases due to the swarming growth characteristics of bacteria within the colony [40]. Singh et al. showed the real-time detection and identification of *Salmonella* colonies grown from inoculated peanut butter, chicken breast, and spinach or from naturally contaminated meat using BARDOT technology (90–100% accuracy) in the presence of background microbiota from naturally contaminated meat [52].

5. Multimode hyperspectral imaging system

Due to the multicomponent nature of foods, their reflectance or fluorescence spectra are complex and chemometric methods using multivariate analysis are needed to extract contaminant-specific information. By varying both the excitation and detection wavelengths and measuring both reflectance and fluorescence emission properties of a food sample, we can fine-tune algorithms for specific foods and contaminants. It has been shown that for biological tissues, dual or multiple excitation fluorescence can increase the specificity and accuracy of classification and quantification of specific sources of fluorescence [54]. Rasch et al. [57] showed the combination of different spectroscopic methods (such as fluorescence and NIR spectroscopy) becomes a promising approach to circumvent such single method inherent limitations and to use optical sensing for in situ mycotoxin detection. Additional chemometric tools are essential to eliminate disturbing factors and to extract the desired biochemical information with respect to contamination with fungi and/or mycotoxins.

An example of a multimode hyperspectral imaging system operates in fluorescence and reflectance modes as well as speckle imaging is shown in **Figure 6** developed by SafetySpect Inc. The system uses spectral band sequential imaging on the detection side. To ensure high signal-to-noise level, camera and spectral selection filter integration time is optimized for each spectral band from visible to the near infrared. The illumination module uses two independent light sources to provide illumination for fluorescence excitation and reflectance measurements using three computer-controlled LED illumination rings. The UVA (375 nm) and blue/violet (420 nm) LED rings provide fluorescence excitation. White LEDs will be used for reflectance illumination. The HSi-440CO hyperspectral imaging system (Gooch & Housego, UK, originally developed by ChromoDynamics, Inc.) incorporated in the proposed system can image and analyze multiple signals in fixed and living cells at video rates. Its tunable filter can switch wavelengths within microseconds. The system acquires multiwavelength, high-spatial and spectral resolution image datasets, and can compute and display quantitative signal-specific images in near real time. The spectrally controllable image capture system can record spectral images of food samples in wavelengths ranging from 450 to 800 nm. The system is configured as a tabletop platform where illumination and detection will operate above the food sample.

In this system, time-varying speckle signals can be quantitatively addressed with the speckle correlation time. A sample containing living microorganisms will have a correlation time way shorter than a static one, and thus contaminated food will be less time-correlated as compared to fresh food due to the spontaneous motility of microorganisms. Correlation time of scattered light from samples, the presence and activity of microorganisms can be quantitatively analyzed.

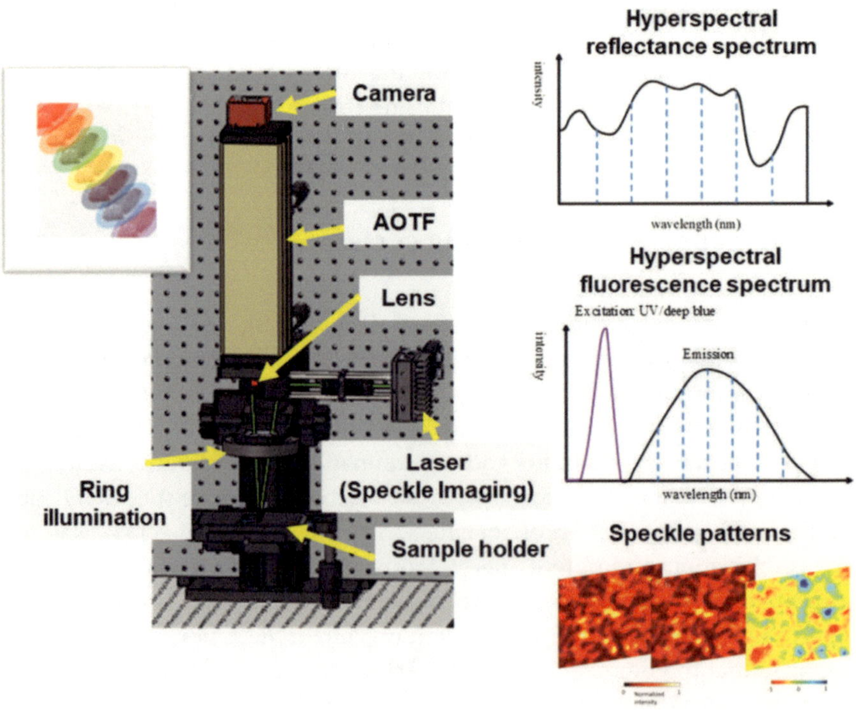

Figure 6. Configuration of the multimode HSI.

Let us consider I (x,y,t) the image of the sample at time t. The correlation coefficient between two images of the sample at different times is given by the normalized autocorrelation function:

$$C(x, y, t) = \frac{1}{T-\tau} \sum_{t=1}^{T-\tau} I(x, y, t) \, . \, I(x, y, t+\tau) \delta t \quad (1)$$

where T is the total acquisition time, δt the time difference, and τ the time lag. In the case of food contamination assessment, the sample is expected to be static and the correlation to be close to the unity. Every decorrelation effect is due, then, to the presence of live microorganisms moving across the sample.

6. Conclusions

There is inherent risk in food (preparing, selling, and consuming it), and we need better ways of minimizing such risk. The number of people who are sickened by problematic food is staggering (it is estimated that 1/6 of the US population is thus affected yearly), and the number of people who die (~3000/year) is unacceptable. If one examines the rather extensive risk management/mitigation literature, it is evident that certain fields of human endeavor (such as air travel) are doing a better job than others in minimizing the undesirable scenarios. A particularly pragmatic take on this field was provided by Dr. J. Reason [58], who developed an approach he termed the Swiss Cheese theory (**Figure 7**). Basically, he posits that we all want to insert countermeasures between us and hazards, to prevent harm, but because we are human and thus imperfect, these countermeasures are like a slice of Swiss cheese. The most logical and direct improvement is to "stack" the slices of cheese, as the holes do not align, and prevention is achieved. Translated to imaging for food safety, this calls for a multimode approach, which is what we propose (see **Figure 8**). The number of modes needed for good performance scales, naturally, with the difficulty of the problem, and we plan to have our implementations reflect this, in the sense that a large, real-time production-level device will be based on more modes than a mid-level (e.g., restaurant) one, while a handheld, portable unit may only address 80% of the challenges, but with ~20% of the cost and size.

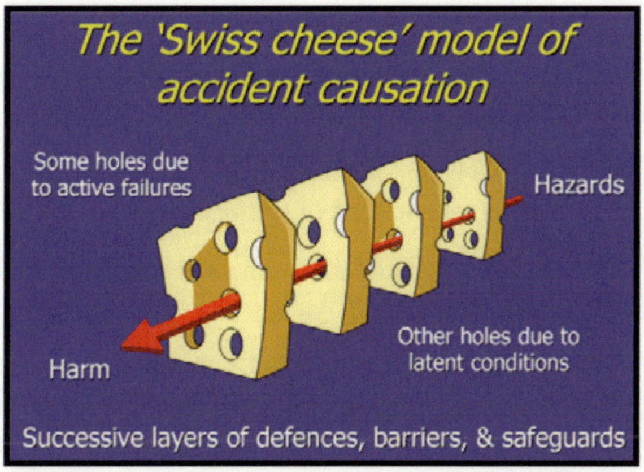

Figure 7. Dr. Reason's Swiss cheese theory of accident causation/prevention.

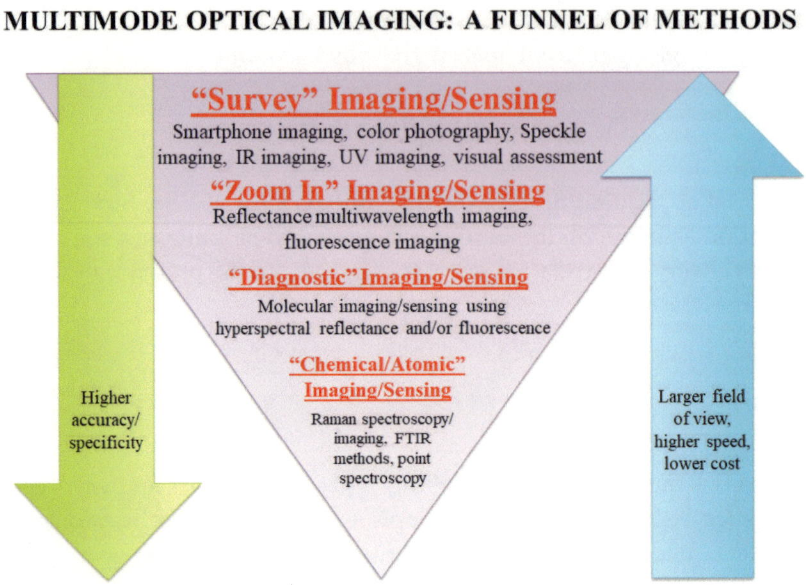

Figure 8. Multimode imaging as a funnel of methods. The right mix (on the same instrument, in the proper sequence) optimizes performance, speed and cost simultaneously.

Author details

Fartash Vasefi, Nicholas Booth, Hesam Hafizi and Daniel L. Farkas*

*Address all correspondence to: dlfarkas@gmail.com

SafetySpect, Inc., Sherman Oaks, CA, USA

References

[1] Gowen AA, O'Donnell C, Cullen PJ, Downey G, Frias JM. Hyperspectral imaging–An emerging process analytical tool for food quality and safety control. Trends in Food Science & Technology. 2007;**18**(12):590-598

[2] Kim MS, Chen YR, Mehl PM. Hyperspectral reflectance and fluorescence imaging system for food quality and safety. Transactions of the ASAE. 2001;**44**(3):721

[3] Goetz AF, Vane G, Solomon JE, Rock BN. Imaging spectrometry for earth remote sensing. Science. 1985 Jun 7;**228**(4704):1147-1153

[4] Liu Y, Chen YR, Kim MS, Chan DE, Lefcourt AM. Development of simple algorithms for the detection of fecal contaminants on apples from visible/near infrared hyperspectral reflectance imaging. Journal of Food Engineering. 2007 Jul 1;**81**(2):412-418

[5] Mehl PM, Chen YR, Kim MS, Chan DE. Development of hyperspectral imaging technique for the detection of apple surface defects and contaminations. Journal of Food Engineering. 2004 Jan 1;**61**(1):67-81

[6] Nicolai BM, Lötze E, Peirs A, Scheerlinck N, Theron KI. Non-destructive measurement of bitter pit in apple fruit using NIR hyperspectral imaging. Postharvest Biology and Technology. 2006 Apr 1;**40**(1):1-6

[7] Xing J, Bravo C, Jancsók PT, Ramon H, De Baerdemaeker J. Detecting bruises on 'golden delicious' apples using hyperspectral imaging with multiple wavebands. Biosystems Engineering. 2005 Jan 1;**90**(1):27-36

[8] Xing J, Jancsók P, De Baerdemaeker J. Stem-end/calyx identification on apples using contour analysis in multispectral images. Biosystems Engineering. 2007 Feb 1;**96**(2):231-237

[9] Xing J, Saeys W, De Baerdemaeker J. Combination of chemometric tools and image processing for bruise detection on apples. Computers and Electronics in Agriculture. 2007 Mar 1;**56**(1):1-3

[10] Weinstock BA, Janni J, Hagen L, Wright S. Prediction of oil and oleic acid concentrations in individual corn (Zea mays L.) kernels using near-infrared reflectance hyperspectral imaging and multivariate analysis. Applied Spectroscopy. 2006 Jan 1;**60**(1):9-16

[11] Ariana DP, Lu R, Guyer DE. Near-infrared hyperspectral reflectance imaging for detection of bruises on pickling cucumbers. Computers and Electronics in Agriculture. 2006 Aug 1;**53**(1):60-70

[12] Cheng X, Chen YR, Tao Y, Wang CY, Kim MS, Lefcourt AM. A novel integrated PCA and FLD method on hyperspectral image feature extraction for cucumber chilling damage inspection. Transactions of the ASAE. 2004;**47**(4):1313

[13] Liu Y, Chen YR, Wang CY, Chan DE, Kim MS. Development of a simple algorithm for the detection of chilling injury in cucumbers from visible/near-infrared hyperspectral imaging. Applied Spectroscopy. 2005 Jan;**59**(1):78-85

[14] Menesatti P, Urbani G, Lanza G. Spectral imaging Vis-NIR system to forecast the chilling injury onset on citrus fruits. InV International Postharvest Symposium. 2004 Jun 6;**682**:1347-1354

[15] Monteiro ST, Minekawa Y, Kosugi Y, Akazawa T, Oda K. Prediction of sweetness and amino acid content in soybean crops from hyperspectral imagery. ISPRS Journal of Photogrammetry and Remote Sensing. 2007 May 1;**62**(1):2-12

[16] Lu R, Peng Y. Hyperspectral scattering for assessing peach fruit firmness. Biosystems Engineering. 2006 Feb 1;**93**(2):161-171

[17] Qiao J, Wang N, Ngadi MO. Water content and weight estimation for potatoes using hyperspectral imaging. In 2005 ASAE Annual Meeting 2005 (p. 1). American Society of Agricultural and Biological Engineers. p. 053126

[18] ElMasry G, Wang N, ElSayed A, Ngadi M. Hyperspectral imaging for nondestructive determination of some quality attributes for strawberry. Journal of Food Engineering. 2007 Jul 1;**81**(1):98-107

[19] Tallada JG, Nagata M, Kobayashi T. Non-destructive estimation of firmness of strawberries (Fragaria× ananassa Duch.) using NIR hyperspectral imaging. Environmental Control in Biology. 2006;**44**(4):245-255

[20] Nagata M, Tallada JG, Kobayashi T. Bruise detection using NIR hyperspectral imaging for strawberry (Fragaria× ananassa Duch.). Environmental Control in Biology. 2006;**44**(2):133-142

[21] Lawrence KC, Windham WR, Park B, Heitschmidt GW, Smith DP, Feldner P. Partial least squares regression of hyperspectral images for contaminant detection on poultry carcasses. Journal of Near Infrared Spectroscopy. 2006 Aug 1;**14**(4):223-230

[22] Park B, Lawrence KC, Windham WR, Smith DP. Performance of hyperspectral imaging system for poultry surface fecal contaminant detection. Journal of Food Engineering. 2006 Aug 1;**75**(3):340-348

[23] Park B, Windham WR, Lawrence KC, Smith DP. Contaminant classification of poultry hyperspectral imagery using a spectral angle mapper algorithm. Biosystems Engineering. 2007 Mar 1;**96**(3):323-333

[24] Kim I, Kim MS, Chen YR, Kong SG. Detection of skin tumors on chicken carcasses using hyperspectral fluorescence imaging. Transactions of the ASAE. 2004;**47**(5):1785

[25] Jiang H, Yoon SC, Zhuang H, Wang W, Lawrence KC, Yang Y. Tenderness classification of fresh broiler breast fillets using visible and near-infrared hyperspectral imaging. Meat Science. 2018 Jan 31;**139**:82-90

[26] Qiao J, Ngadi MO, Wang N, Gariépy C, Prasher SO. Pork quality and marbling level assessment using a hyperspectral imaging system. Journal of Food Engineering. 2007 Nov 1;**83**(1):10-16

[27] Zhu WY, Su WH, Sun DW. Measurement of tenderness of red meats using hyperspectral imaging: A brief review. Biosystems and Food Engineering Research Review. 2017 May;**22**:66

[28] Xiong Z, Sun DW, Zeng XA, Xie A. Recent developments of hyperspectral imaging systems and their applications in detecting quality attributes of red meats: A review. Journal of Food Engineering. 2014 Jul 1;**132**:1-3

[29] Qin J, Chao K, Kim MS. Raman chemical imaging system for food safety and quality inspection. Transactions of the ASABE. 2010;**53**(6):1873-1882

[30] Bianco V, Mandracchia B, Nazzaro F, Marchesano V, Gennari O, Paturzo M,Ferraro P, et. al Food quality inspection by speckle decorrelation properties of bacteria colonies. In: Optical Methods for Inspection, Characterization, and Imaging of Biomaterials III (Vol. 10333, p. 103331N). International Society for Optics and Photonics; 2017, June

[31] Yoon J, Lee K, Park Y. A simple and rapid method for detecting living microorganisms in food using laser speckle decorrelation. arXiv preprint arXiv:1603.07343; 2016

[32] Kamruzzaman M, ElMasry G, Sun DW, Allen P. Non-destructive prediction and visualization of chemical composition in lamb meat using NIR hyperspectral imaging and multivariate regression. Innovative Food Science & Emerging Technologies. 2012;**16**:218-226

[33] Tao F, Peng Y. A method for nondestructive prediction of pork meat quality and safety attributes by hyperspectral imaging technique. Journal of Food Engineering. 2014;**126**:98-106

[34] Pu YY, Feng YZ, Sun DW. Recent progress of hyperspectral imaging on quality and safety inspection of fruits and vegetables: A review. Comprehensive Reviews in Food Science and Food Safety. 2015;**14**(2):176-188

[35] Cheng JH, Sun DW. Rapid quantification analysis and visualization of *Escherichia coli* loads in grass carp fish flesh by hyperspectral imaging method. Food and Bioprocess Technology. 2015;**8**(5):951-959

[36] Feng YZ, ElMasry G, Sun DW, Scannell AG, Walsh D, Morcy N. Near-infrared hyperspectral imaging and partial least squares regression for rapid and reagentless determination of Enterobacteriaceae on chicken fillets. Food Chemistry. 2013;**138**(2): 1829-1836

[37] Del Fiore A, Reverberi M, Ricelli A, Pinzari F, Serranti S, Fabbri AA,Fanelli C, et al. Early detection of toxigenic fungi on maize by hyperspectral imaging analysis. International Journal of Food Microbiology. 2010;**144**(1):64-71

[38] Fan C, Hu Z, Mustapha A, Lin M. Rapid detection of food-and waterborne bacteria using surface-enhanced Raman spectroscopy coupled with silver nanosubstrates. Applied Microbiology and Biotechnology. 2011;**92**(5):1053-1061

[39] Sowoidnich K, Schmidt H, Kronfeldt HD, Schwägele F. A portable 671 nm Raman sensor system for rapid meat spoilage identification. Vibrational Spectroscopy. 2012;**62**:70-76

[40] Kim H, Singh AK, Bhunia AK, Bae E. Laser-induced speckle scatter patterns in Bacillus colonies. Frontiers in Microbiology. 2014;**5**:537

[41] Peng Y, Zhang J, Wang W, Li Y, Wu J, Huang H, Jiang W, et al. Potential prediction of the microbial spoilage of beef using spatially resolved hyperspectral scattering profiles. Journal of Food Engineering. 2011;**102**(2):163-169

[42] Gómez-Sanchis J, Moltó E, Gomez-Chova L, Aleixos N, Camps-Valls G, Juste F, Blasco J. Hyperspectral computer vision system for the detection of Penicillium digitatum in citrus packing lines. In: 2004 CIGR International Conference, Beijing, China. 2004, October. (pp. 11-14)

[43] Barbin DF, ElMasry G, Sun DW, Allen P, Morsy N. Non-destructive assessment of microbial contamination in porcine meat using NIR hyperspectral imaging. Innovative Food Science & Emerging Technologies. 2013;**17**:180-191

[44] Lohumi S, Lee S, Lee H, Cho BK. A review of vibrational spectroscopic techniques for the detection of food authenticity and adulteration. Trends in Food Science & Technology. 2015 Nov;**46**(1, 1):85-98

[45] Kim MS, Lefcourt AM, Chao K, Chen YR, Kim I, Chan DE. Multispectral detection of fecal contamination on apples based on hyperspectral imagery: Part I. Application of visible and near–infrared reflectance imaging. Transactions of the ASAE. 2002;**45**(6):2027

[46] Vargas AM, Kim MS, Tao Y, Lefcourt AM, Chen YR, Luo Y, et al. Detection of fecal contamination on cantaloupes using hyperspectral fluorescence imagery. Journal of Food Science. 2005;**70**(8):471-476

[47] Yang CC, Jun W, Kim MS, Chao K, Kang S, Chan DE, Lefcourt A. Classification of fecal contamination on leafy greens by hyperspectral imaging. Sensing for Agriculture and Food Quality and Safety II, Proceedings of the SPIE. 2010;**7676**:76760F-767601F

[48] Domingo E, Tirelli AA, Nunes CA, Guerreiro MC, Pinto SM. Melamine detection in milk using vibrational spectroscopy and chemometrics analysis: A review. Food Research International. 2014 Jun 1;**60**:131-139

[49] Zhou N, Zhou Q, Meng G, Huang Z, Ke Y, Liu J, Wu N. Incorporation of a basil-seed-based surface enhanced Raman scattering sensor with a pipet for detection of melamine. ACS Sensors. 2016 Oct 14;**1**(10):1193-1197

[50] Li X, Feng S, Hu Y, Sheng W, Zhang Y, Yuan S, Zeng H, Wang S, Lu X. Rapid detection of melamine in milk using immunological separation and surface enhanced Raman spectroscopy. Journal of Food Science. 2015 Jun 1;**80**(6):C1196–C1201

[51] Everard CD, Kim MS, Cho H, O'Donnell CP. Hyperspectral fluorescence imaging using violet LEDs as excitation sources for fecal matter contaminate identification on spinach leaves. Journal of Food Measurement and Characterization. 2016 Mar;**10**(1, 1):56-63

[52] Singh AK, Bettasso AM, Bae E, Rajwa B, Dundar MM, Forster MD, Liu L, Barrett B, Lovchik J, Robinson JP, Hirleman ED. Laser optical sensor, a label-free on-plate salmonella enterica colony detection tool. MBio. 2014 Feb 28;**5**(1):e01019-e01013

[53] Yoon J, Lee K, Park Y. A simple and rapid method for detecting living microorganisms in food using laser speckle decorrelation. arXiv preprint arXiv. 2016 Mar 18;**1603**:07343

[54] Sauvage VR, Levene AP, Nguyen HT, Wood TC, Kudo H, Concas D, Thomas HC, Thursz MR, Goldin RD, Anstee QM, Elson DS. Multi-excitation fluorescence spectroscopy for analysis of non-alcoholic fatty liver disease. Lasers in Surgery and Medicine. 2011;**43**(5):392-400

[55] Kim H, Singh AK, Bhunia AK, Bae E. Laser-induced speckle scatter patterns in bacillus colonies. Frontiers in Microbiology. 2014 Oct 14;**5**:537

[56] Qin J, Chao K, Kim MS. Development of a Raman chemical imaging system for food safety inspection. In 2010 Pittsburgh, Pennsylvania, June 20-June 23, 2010 2010 (p. 1). American Society of Agricultural and Biological Engineers

[57] Rasch C, Kumke M, Löhmannsröben HG. Sensing of mycotoxin producing fungi in the processing of grains. Food and Bioprocess Technology. 2010 Dec 1;**3**(6):908-916

[58] Reason J. Human error: models and management. BMJ. 2000;**320**(7237):768-770

Chapter 3

Hyperspectral Imaging and Their Applications in the Nondestructive Quality Assessment of Fruits and Vegetables

Xiaona Li, Ruolan Li, Mengyu Wang, Yaru Liu,
Baohua Zhang and Jun Zhou

Additional information is available at the end of the chapter

http://dx.doi.org/10.5772/intechopen.72250

Abstract

Over the past decade, hyperspectral imaging has been rapidly developing and widely used as an emerging scientific tool in nondestructive fruit and vegetable quality assessment. Hyperspectral imaging technique integrates both the imaging and spectroscopic techniques into one system, and it can acquire a set of monochromatic images at almost continuous hundreds of thousands of wavelengths. Many researches based on spatial image and/or spectral image processing and analysis have been published proposing the use of hyperspectral imaging technique in the field of quality assessment of fruits and vegetables. This chapter presents a detailed overview of the introduction, latest developments and applications of hyperspectral imaging in the nondestructive assessment of fruits and vegetables. Additionally, the principal components, basic theories, and corresponding processing and analytical methods are also reported in this chapter.

Keywords: hyperspectral imaging, fruits and vegetables, nondestructive quality assessment

1. Introduction

In recent years, consumer demand for fruits and vegetables tends to be diversified, and more attention has been paid to the external quality of apples. Generally, such attributes include its ripeness, size, weight, shape, color, condition, or presence/absence of defects, stems or seeds, as well as a series of internal properties such as sweetness, acidity, texture, hardness, among others [1]. Consequently, the accurate, rapid, and objective assessment system in the processing stage is essential to ensure the quality of fruits and vegetables during processing operations. Food process control necessitates real-time monitoring at critical processing points [2].

Traditional optical sensing techniques, such as imaging and spectroscopy, have limitations to acquire adequate spatial and spectral information for nondestructive evaluation of food and agricultural products. Generally, conventional imaging cannot acquire spectral information and spectroscopy measurement cannot cover large sample area. In general, the frequently-used vision systems for fruits and vegetables sorting are based on color video camera that imitates the vision of the human eye by capturing images using three filters centered on red, green and blue (RGB) wavelengths [3, 4]. Thus, they are limited to observing scenes and are usually not able to obtain much information about the external or internal composition of the products or to detect some defects or alteration whose color is similar to the color of the sound skin. In addition, traditional methods of fruits and vegetables monitoring involving analytical techniques are too time consuming, expensive and require sample destruction.

Over the past decades, with the rapid development of information science, image processing and pattern recognition technology, optical sensing technologies have been emerged as scientific tools for nondestructive assessment for quality of fruits and vegetables. Spectral imaging technology, combining conventional imaging and spectroscopy techniques, can acquire spatial and spectral information from the target, which is used for evaluating individual food products. In particular, hyperspectral imaging has been widely researched and developed by integrating spectroscopy and imaging techniques into a system that can obtain a spatial map of spectral variation, resulting in many successful applications in the quality assessment of fruits and vegetables. A typical spectral image is composed of a set of monochromatic images corresponding to certain wavelengths, and hyperspectral image systems have the natural advantage compared to the traditional computer vision, even the human vision [2]. Hyperspectral imaging systems can make it possible to extract some appearance features that are difficult or impossible with the traditional computer vision systems.

This chapter focuses on hyperspectral imaging technologies in the quality nondestructive assessment of fruits and vegetables. In the second section, overview, components, and different image acquisition technologies of hyperspectral imaging are explained and discussed. Hyperspectral images generate a large amount of information that can be processed using different statistical techniques [1]. In the third section, varying nondestructive processing and analysis methods are illustrated in detail. Finally, applications of this technology are discussed, and conclusions are given.

2. Hyperspectral imaging technique

2.1. Overview of hyperspectral imaging

Hyperspectral imaging, known also as chemical or spectroscopic imaging, is an emerging technique that integrates conventional imaging and spectroscopy to simultaneously collect spatial and spectral information from an object. The term "hyperspectral imaging" was derived from works in remote sensing first mentioned by Goetz et al. in [5] to make a direct identification of surface materials in the form of images. Although originally developed for remote sensing, hyperspectral imaging system is gradually found to have natural advantages over the traditional computer vision systems [2] in such diverse fields as agriculture [6–9]. With the development of optical sensing and imaging techniques, hyperspectral imaging has recently emerged as a scientific and efficient inspection and assessment tool for quality of fruits and vegetables. The goal of hyperspectral imaging is to obtain the spectrum for each pixel in the image of a scene, with the purpose of finding objects, identifying materials, or detecting processes [10]. To obtain high spectral resolution and narrow band image data, hyperspectral imaging is generally combined with spectroscopic technique, two-dimensional geometric space and one-dimensional spectral information detection.

2.2. Components of hyperspectral imaging system

Figure 1 shows the schematic of the hyperspectral imaging system commonly used in our research. As shown in **Figure 1**, a typical hyperspectral imaging system usually consists of the following components: a light source (illumination), a wavelength dispersion device (spectrograph), an area detector (camera), a transportation stage and a computer with corresponding software [11].

Light source for spectral imaging applications can generally be classified into two categories: illumination and excitation source. Broadband lights are generally used as the illumination sources for reflectance and transmittance imaging while narrowband lights are commonly used as the excitation sources. Therefore, illumination is a crucial part of the hyperspectral imaging system. Compared with the naked eyes, vision systems are affected by the level and quality of illumination. Illumination devices generate light that illuminates the inspected target objects; thus, the performance of the illumination system can greatly influence the quality of images and plays an important role in the overall efficiency and accuracy of the system [12]. Good illumination can help to improve the success of the image processing and analysis by reducing noise, shadow, reflection, and enhancing image contrast [2]. In addition,

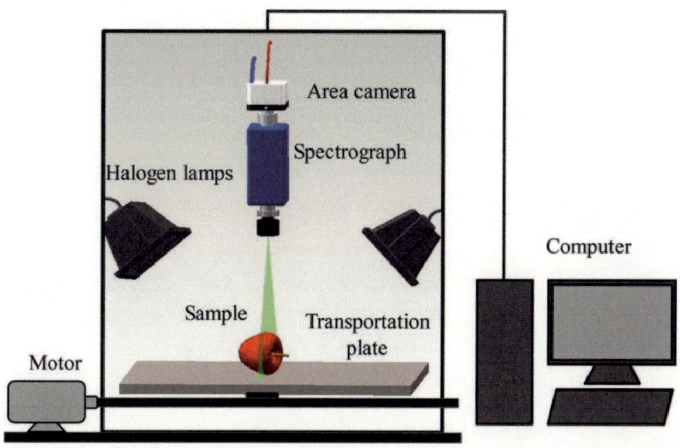

Figure 1. A schematic of the hyperspectral imaging system.

the positions, types of lamps, and color quality of the illumination are all considered when choosing the most suitable illumination. Incandescent lamps, fluorescent lamps, lasers, and infrared lamps are the commonly used light sources [13].

The wavelength dispersion device is one of the key components of hyperspectral imaging system. Filter, grating and prism are three typical wavelength dispersion devices. These optical devices are used to disperse broadband light into different wavelengths and project the scattered light onto the area detector. The principles of prism and diffraction grating are illustrated in **Figure 2**. In a word, filter is always used in the multispectral imaging system, while prism and grating are widely used in the hyperspectral imaging system [2]. Besides, the efficiencies of the transmission components (e.g., prisms) are generally lower than those of the reflective optical component (e.g., mirrors). An optical wavelength dispersion device includes [14, 15]: a first substrate; an input unit formed on the first substrate having a slit for receiving an optical signal; a grating line formed on the first substrate for generating a diffracted light beam of the optical signal; a first optical reflector formed on the first substrate to the reflected output beam from the diffraction grating for the output; and a second substrate covered on the top of the input unit and the grating. The wavelength dispersion is capable to disperse

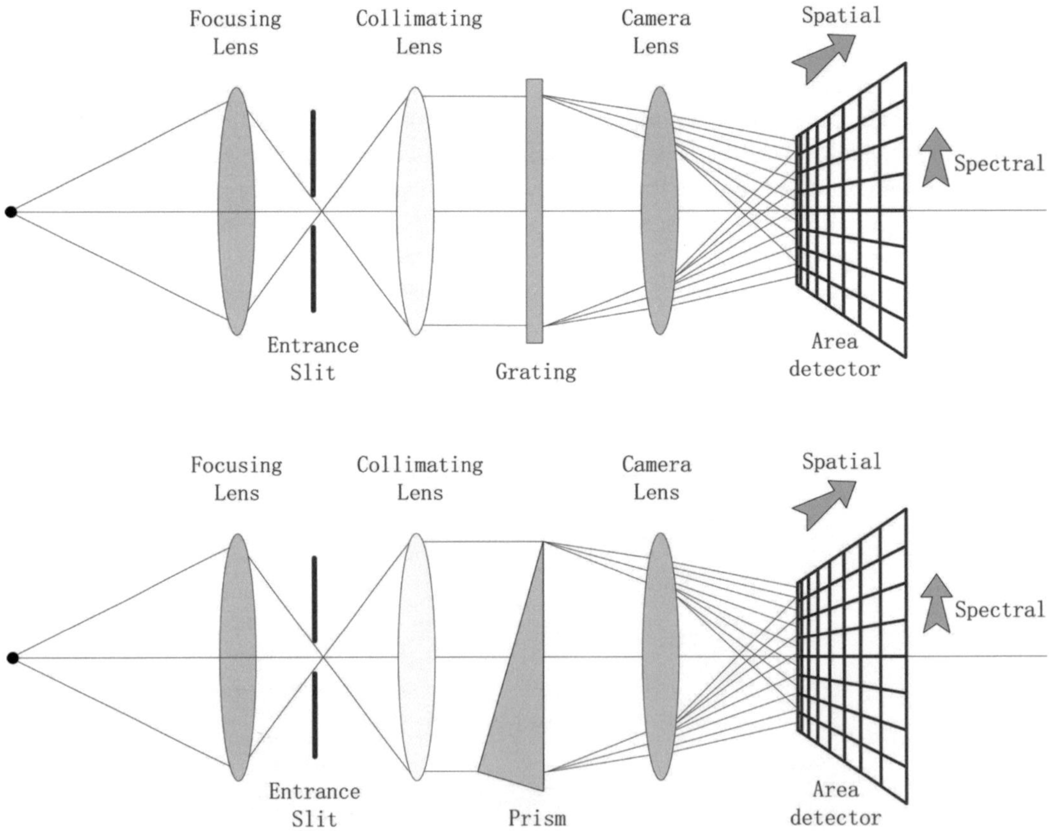

Figure 2. Operating principles of diffraction grating and prism.

broadband light into varying wavelengths. Typical examples include filter wheels, imaging spectrographs, acousto-optic tunable filters, liquid crystal tunable filters, Fourier transform imaging spectrometers, and single shot imagers [16].

The camera, which is one of the image acquisition devices, is another core component of the hyperspectral imaging system. It is the carrier of the physical or chemical information and the light generated from the light source. Other image acquisition devices used in food applications are computed tomography (CT), magnetic resonance imaging (MRI), ultrasound and electrical tomography [17]. Charge coupled device (CCD) and complementary metal oxide semiconductor (CMOS) image sensors are two different means to generate the image digitally [2]. A CCD is a device for the movement of electrical charge, generally from within the device to an area where the charge can be manipulated. In the CCD image sensor, pixels are represented by P-doped metal oxide semiconductor (MOS) capacitors. When image acquisition starts, these capacitors are biased above the threshold for inversion, allowing the conversion of incoming photons into electron charges at the semiconductor-oxide interface [18]; then, the CCD is used to read out these charges. The CMOS image sensor consists of millions of pixel sensors, each of which includes a photo detector. As light enters the camera through the lens, it strikes the CMOS image sensor, allowing each photo detector to accumulate an electric charge based on the amount of light that strikes it. CMOS is also sometimes referred to as complementary-symmetry metal–oxide–semiconductor (COS-MOS). In general, the CMOS image sensor is used in applications with less exacting quality demands, and the CCD image sensor is widely used in medical, scientific and professional applications where high-quality image data are required.

Compared with the traditional computer vision system, a wavelength dispersion device and a transportation stage are additional components of hyperspectral or multispectral computer vision systems. The translation stage is used to move the sample past the objective lens when the camera captures only a line of the illuminated object.

The computer is not only used to control the hyperspectral imaging system for data acquisition, processing and analysis of image and spectral data for specific application, but also can provide storage space for hyperspectral image. By scanning the entire surface of the specimen, a complete hyperspectral image is created and displayed by the computer [19].

2.3. Generation of hyperspectral images

Hyperspectral image is three-dimensional hyperspectral cube, composed of two spatial and one wavelength dimension [20]. There are three approaches to build hyperspectral images based on the method by which spatial information is acquired as whiskbroom, pushbroom, and tunable filter known as point scanning, line scanning, and area scanning, respectively [21], as illustrated in **Figure 3**. The point-scan method (**Figure 3a**) is a basic spectroscopic approach, where a single point is scanned along two spatial dimensions by moving the sample or the detector. When a single point is scanned, the sample moves to the next measurement point and another spectrum is captured. By moving the sample systematically in two spatial dimensions, a complete hyperspectral image can be obtained. However, it is not suited for fast image acquisition because the scan of many points for two spatial dimensions is a time-consuming process. The line-scan method (**Figure 3b**) can be considered as an extension of point scanning method. In the line-scan

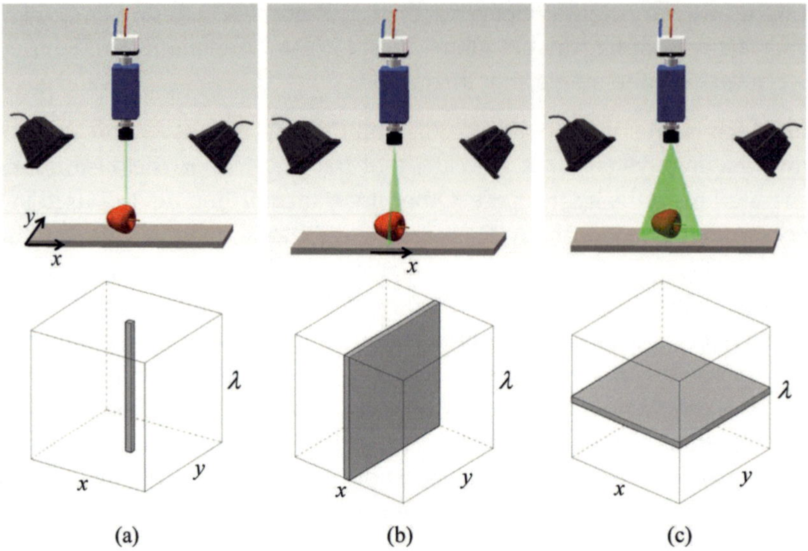

Figure 3. Three different approaches to generate a hyperspectral image. (a) The point-scan method. (b) The line-scan method. (c) The line-scan method.

method, a slit of spatial information and full spectral information for each spatial point in the linear field of view can be acquired simultaneously. But the line-scan method requires the use of an imaging spectrometer, in which a diffraction grating disperses light entering through a thin slit and projects. Food commodities normally are moved linearly along a production line [11]. Consequently, the line-scan method is appropriate for online inspection of individual food. The area-scan method (**Figure 3c**) does not require the relative movement between the sample and the detector and is usually used to collect images from the fixed scene. The line-san camera holds an advantage over area-scan camera. Unlike these area-scan cameras, a line-scan camera can expose a new image while the previous image is still reading out its data. A detailed description of data preprocessing methods can be found in the literature [22, 23].

As shown in **Figure 4**, hyperspectral imaging system is generally carried out in reflectance, transmittance or interactance modes according to the specific light-output captured by hyperspectral imaging system [24]. In the external quality inspection of fruits and vegetables, the reflectance mode is considered to be the most suitable approach. Position of light source and the optical detector (cameral, spectrograph, and lens) are different for each acquisition mode [21]. In the external quality inspection of fruits and vegetables, the reflectance mode (**Figure 4a**) is considered to be the most suitable approach. In reflectance mode, to avoid specular reflection, the detector captured the reflected light from the illuminated sample in a specific conformation. The transmitted light captured through the sample is often very weak but carries more valuable information and the detector is located on the opposite side of the light source. Transmittance mode (**Figure 4b**) is usually used to determine internal component concentration and detect internal defects of relative transparent materials [16]. Interactance mode (**Figure 4c**) is a combination of reflectance and transmittance where both light source and the detector are located in the same side of sample and parallel to each other.

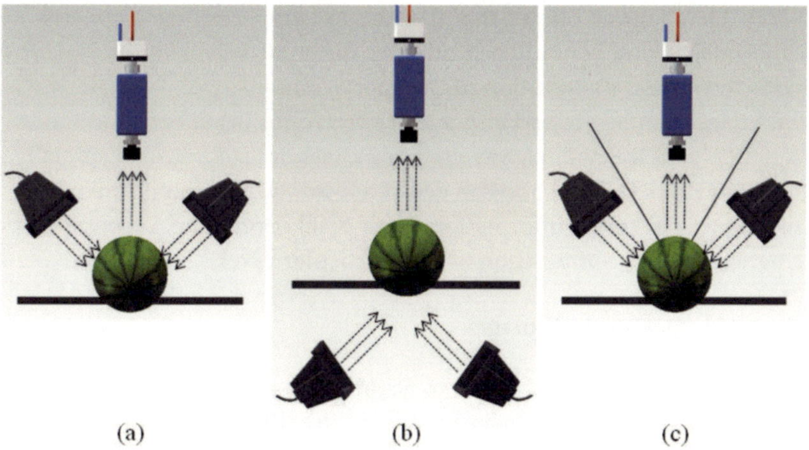

Figure 4. Three different modes to generate a hyperspectral image. (a) The reflectance mode. (b) The transmittance mode. (c) The interactance mode.

2.4. Characteristics of the hyperspectral images

In the conventional RGB images, some unobvious quality character, which is even not visible to the human eyes, is impossible or difficult to detect. Unlike the conventional RGB images, whose spectrum information is very limited, the hyperspectral images contain extensive monochromatic image [2]. In one or several monochromatic images, the unobvious external quality characters can be very clear or easy to detect. Hyperspectral images are composed of numerous continuous wavebands for spatial position of an object studied.

Figure 5 illustrates the conceptual view of a hyperspectral image, which contains a stack of two-dimensional images one behind each other at different wavelengths and can be described

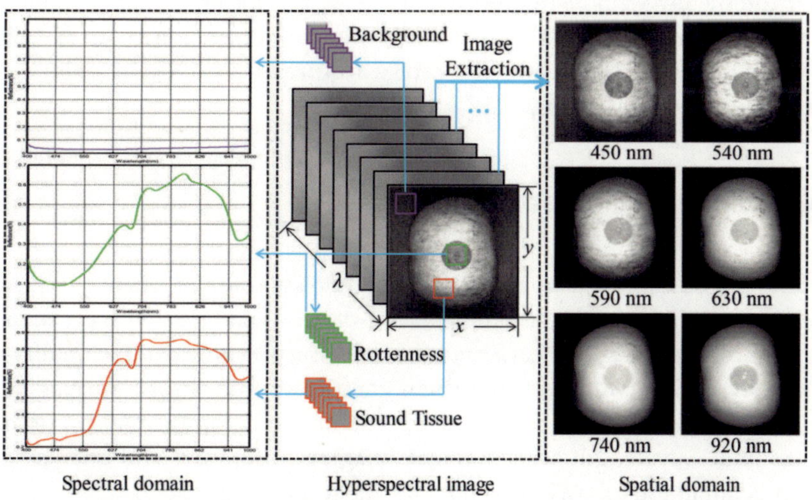

Figure 5. The conceptual view of a hyperspectral image with spectral and spatial domains.

as I(x, y, λ) [21]. The diagram shows that the raw hyperspectral cube consists of a series of contiguous sub-images one behind each other at different wavelengths [16], and each sub-image provides the spatial distribution of the spectral intensity at a certain wavelength. The hyperspectral images can be viewed either as a spectrum I (λ) at each individual pixel (x, y) or as an image I (x, y) at individual wavelength λ. Each image acquires spatially distributed spectral information at pixel level and can be used to analyze the biochemical constituent of a sample according to the spatial information. Each pixel containing a complete spectrum can be used to characterize the composition of that particular pixel.

2.5. Calibration of hyperspectral images

The hyperspectral imaging is a useful tool to acquire and record the raw hyperspectral information of fruits and vegetables. However, due to the differences in camera quantum and physical configuration of imaging systems, the uncorrected radiance for the different systems, even for the same system used in different times, might be very different for the same sample taken under the same condition [25]. Therefore, accurate calibrations for a hyperspectral imaging system are necessary to guarantee the stability and acceptability of the extracted hyperspectral image data and the consistent performance of the system. The original hyperspectral images can be calibrated into the reflectance mode based on black and white reference images. The hyperspectral reflectance images R for a spatial pixel (i) at a given wavelength was calculated by using the following equation [26, 27].

$$R_i = \left(\frac{RS_i - RD_i}{RW_i - RD_i}\right) \times 100\% \quad (1)$$

where *RS*, *RD*, and *RW* are respectively the raw intensity values of identical pixels from the sample image, dark reference image, and white reference images. R_i is the calibrated hyperspectral image in a unit of relative reflectance. The dark reference image *RD* (with ~0% reflectance), which can be obtained with the light source turned off completely and the camera lens covered completely with its nonreflective opaque black cap, is used to remove the dark current effect of the area detectors [28]. The white reference image *RW* (with ~99% reflectance) represents the highest intensity values. *RW* can be acquired from a Teflon white surface under the same condition of the raw image.

3. Nondestructive assessment methods

The spectrum may be complicated by instrumental noise, complex chemical composition of products, environmental factors and other sources of variability [19]. As a consequence, spectral and image preprocessing and correction are necessary to improve the quality of the data before data analysis [29]. Moreover, the chemometrics is crucial for information extraction and better interpretation of the acquired data. The methods for spectral preprocessing and correction, optimal wavelength selection, and imaging processing and analysis models are introduced in detail in the following sections, as illustrated in **Figure 6**.

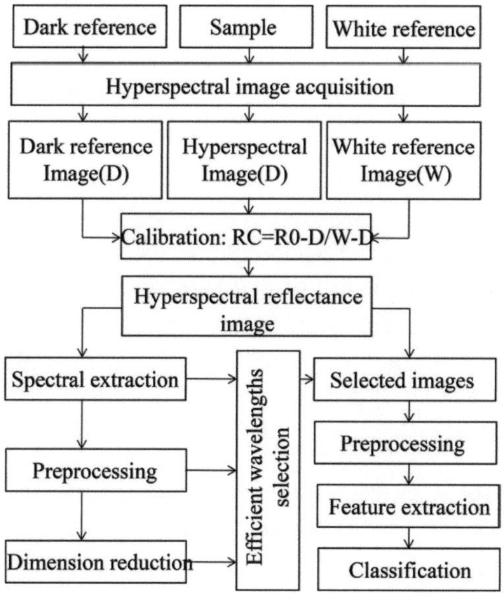

Figure 6. The flowchart of analyzing methods with hyperspectral image.

3.1. Spectral analysis methods

3.1.1. Spectral preprocessing methods

The spectra of solid and scattering samples such as vegetables are influenced by physical properties such as shape, size, etc. This creates baseline shifts and noises in the spectra with broad wavelength regions when analyzing quality parameters [30]; thus, preprocessing of near-infrared (NIR) spectral data has become an integral part of chemometrics modeling. The goal of the preprocessing is to remove physical effects in the spectra in order to improve the subsequent multivariate regression, classification model or exploratory analysis. Selecting suitable preprocessing methods should always be considered in relation to the successive modeling stage. The whole data processing generally consists of the following several steps: spectral preprocessing, calibration model and model validation. A detailed description of data preprocessing methods can be found elsewhere [24, 31]. Some of the preprocessing methods are presented in the following sections.

3.1.1.1. Averaging

Averaging over spectra is generally performed during the acquisition spectrum to reduce the thermal noise of the detector. The number of scans depends on the application: the PDA spectrophotometer operates at a typical acquisition time of less than 50 ms, with almost no time to get multiple scans in the online classification, while the PDA spectrophotometer measurement time is less critical and can average several spectra without affecting the measurement throughput rate in the laboratory [32]. Averaging over wavelengths is used to smooth the spectrum or to reduce the number of wavelengths. Overall, most spectrophotometers may provide a better spectral resolution than the actual optical resolution.

3.1.1.2. Centering

For all practical purposes, it is recommended that data be centered or mean centered. The first stage in centering is often to subtract the average from each variable. The objective of centering is to ensure that all results will be interpretable in term of variation around the mean [32]. This is especially crucial if the variables differ significantly in their relative magnitudes, as the values with the greatest variance will be favored in regression analysis.

3.1.1.3. Smoothing

Smoothing is used to reduce high-frequency noise from the spectral data and signal-to-noise without reducing the number of spectral variables. Its principle is to acquire an optimal estimation value by averaging or fitting several points in a window. The broader the window is, the lower the spectral resolution would be [24]. Consequently, it is important to choose the window width properly. Smoothing improves the vision of the original spectra in addition to remove the useless information. Based on different smoothing fit methods, smoothing could be divided into moving average smoothing, Gaussian filter smoothing, median filter smoothing and Savitzky-Golay smoothing (S-G smoothing) [33, 34]. Different smoothing algorithms are adapted to different specific types of noise models. In other words, the appropriate smoothing algorithm should be selected flexibly according to the noise situation contained in the actual image.

3.1.1.4. Standard normal variate

Standard normal variate (SNV) is a row-oriented transformation which is capable of removing the multiplicative interferences from spectral caused by scatter and particle size effects from spectral data. SNV removes scatter effects by centering and scaling each individual spectrum [35, 36]. The method assumes that the absorbance of each wavelength point in the spectrum meets some certain distribution such as Gaussian distribution. Each spectrum can be calibrated based on this assumption. Firstly, the average value of a spectrum is subtracted from the original spectrum, and then the result is divided by the standard deviation [24]. This method is widely used when the variables are measured in different ranges or in different units, and it cannot be used for NIR spectroscopy because the noise from variables with small standard deviations may explode and lead to unreliable or incorrect models.

3.1.1.5. Multiplicative scatter correction

Multiplicative scatter correction (MSC) is a transformation method used to compensate for additive or multiplicative effects in spectral data [36, 37]. It is performed by correcting the scatter level of each to the level of an average spectrum. Similar to SNV, the objective of MSC is to eliminate the deviations caused by particle size and scattering [36]. The difference is that

MSC standardizes every spectrum using the mean spectrum of all spectra while SNC use only the data from that spectrum. Therefore, for MSC effects on each spectrum alone, the correction capability of MSC is usually weaker than that of SNV. In SNV correction, each individual spectrum is normalized to zero mean and unit variance [32].

3.1.1.6. *Derivative correction*

Derivative is used to remove overlapping peaks and baseline shifts induced by the variation of particle sizes and instrumental conditions, so that more details within the spectra can be revealed [31, 32]. The first derivative of a spectrum is simply a measure of the slope of the spectral curve at every point [38, 39]. The slope of the curve is not affected by baseline offsets in the spectrum, and thus, the first derivative is a very effective method for removing baseline offsets. However, peaks in raw spectra usually become zero-crossing points in first derivative spectra, which can be difficult to interpret. The second derivative is a measure of the change in the slope of the curve. In addition to ignoring the offset, it is not affected by any linear that may exist in the data, and is therefore a very effective method for removing both the baseline offset and slope from a spectrum. The second derivative can help resolve nearby peaks and sharpen spectral features. Peaks in raw spectra change sign and turn to negative peaks with lobes on either side in the second derivative. Two commonly used spectral derivative approaches are Gap-Segment derivative and Savitzky-Golay (S-G) derivative [24].

3.1.1.7. *Transformation*

In spectral analysis, Fourier transformation (FT) and Wavelet transformation (WT) are often used for data compression, smoothing and filtering, as well as for the extraction of effective information. FT is a very important signal processing technique, which can realize the transformation between time domain functions and frequency domain functions. The principle of it is to decompose the original spectrum into the sum of sinusoidal waves of many varying amplitudes, frequencies and directions. WT is based on the idea of decomposing chemical signals into scale compositions according to their different frequencies by applying a basis function [24]. WT is similar to FT with a completely different merit function. The main difference is that FT decomposes the signal into sines and cosines; in contrary, WT uses functions that are localized in both the real and Fourier space [40].

3.1.2. *Optimal wavelength selection methods*

Due to the high resolution of modern spectroscopy instrumentations, the acquired spectral data set may have thousands of variables/wavelengths and hundreds or thousands of samples [41, 42]. Thus the hyperspectral imaging inspection algorithm will be very time-consuming due to the large-scale massive data. In order to simplify the complexity of computation, improve the efficiency of the detection, and meet the inspection speed required by the industry, variable selection (wavelength selection) is the most necessary and important step to select the optimal variables and remove the highly calibrated variables [43]. Many methods based on different criteria have been developed for this purpose. Some of them include competitive adaptive reweighted sampling (CARS), random frog (RF), successive projections algorithm (SPA), genetic algorithm (GA) and uninformative variables elimination (UVE) which can be implemented prior to the construction of both regression and classification models.

3.1.2.1. Competitive adaptive reweighted sampling (CARS)

Competitive adaptive reweighted sampling (CARS) is a novel wavelength selection algorithm employing the "survival of the fittest" principle from Darwin's evolution theory [44]. It is originally developed to select informative wavelengths from contiguous spectral data, specifically applied for the first time to NIR spectroscopy. The method selects wavelength subsets sequentially from the sampling runs in an iterative manner. It basically consists of a number of iterations involving [45]: (1) Monte Carlo (MC) model sampling, (2) wavelength reduction by exponentially decreasing function (EDF), (3) wavelength reduction by adaptive reweighted sampling (ARS), and (4) model building with each subset of selected variables and CV to calculate prediction error. **Figure 7** shows the scheme of the CARS algorithm. For each MCS run or iteration, the four steps mentioned above will be repeated, obtaining an error for each one. Finally, the subset with the lowest RMSECV value will be determined as the optimal subset [46]. The key wavelengths selected by CARS are considered as the wavelengths with the large absolute regression coefficients in a multivariate linear regression model. The exponential decay function is used to control the retention rate of variable in the algorithm, and it has the potential to select an optimal combination of the wavelengths.

3.1.2.2. Random frog (RF)

Random frog (RF) algorithm is a useful wavelength selection technique based on the framework of reversible jump Markov chain Monte Carlo (MCMC) or the multiple decision

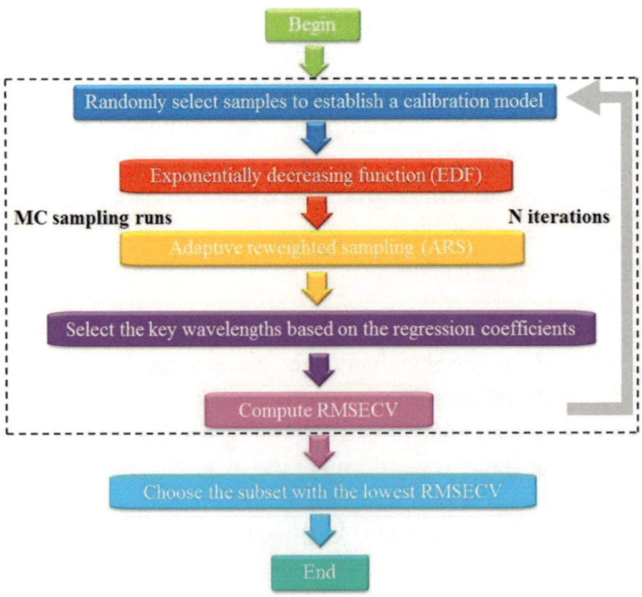

Figure 7. The flowchart of the competitive adaptive reweighted sampling algorithm.

trees. Like CARS, it works in an iterative manner; meanwhile, it calculates the selection probability for each variable. Briefly, random frog works in three steps [47, 48]: (1) initializing randomly a variable subset V_0 containing Q variables; (2) generating a candidate variable subset V^* including Q^* variable; accept V^* as V_1 with a certain probability and let $V_0 = V_1$; repeat the above procedures until N iterations are finished; and (3) computing a selection probability of each variable which can be used as a measure of variable importance. The schematic is shown in **Figure 8**. The advantage of random frog is that it does not require any rigorous mathematical formula. And it do not need to use the previous distribution in formal reversible jump MCMC methods, which makes it easier to implement and compute. There are five tuning parameters to control the RF performance, which can be optimized in the routine. The two most important parameters are the number of variables contained in the number of iterations and the initial variable set.

3.1.2.3. Successive projections algorithm (SPA)

The successive projections algorithm (SPA), a forward selection method which uses simple operations in a vector space to minimize variable collinearity, is a novel variable selection strategy in hyperspectral image analysis for multivariate calibration [49, 50]. The main purpose of SPA is to select wavelengths with minimal redundancy [43]. In summary, the steps to execute SPA are: (1) carrying out projections on the N matrix and generating K chains of M variables each, (2) evaluating candidate subsets of variables extracted from the chains gener-

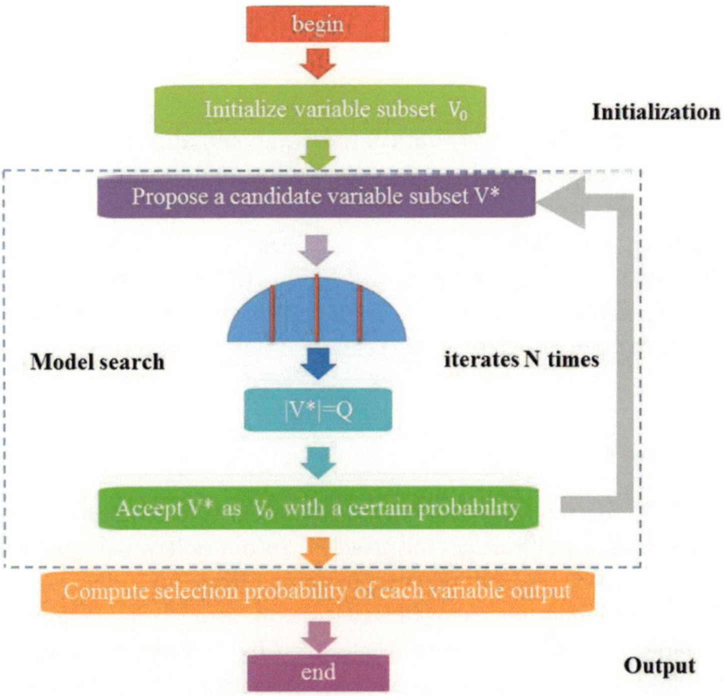

Figure 8. The flowchart of the random frog algorithm.

ated in the first phase, and (3) eliminating procedures aimed at discarding uninformative variables without significant loss of prediction capability. Many successful applications have proven SPA to be an outstanding variable selection approach.

3.1.2.4. Genetic algorithm (GA)

The Genetic algorithm (GA) is an effective globe searching algorithm. Based on a fitness function, GA is an iterative process starting from a population of randomly generated individuals and achieves optimal solutions through genetic operations including crossover, selection and mutation [24]. The steps of GA involved are [51]: (1) building an initial population of variable sets by setting bits for each variable randomly, (2) fitting a PLS regression model to each variable set and computing the performance, (3) a collection of variable sets with higher performance are selected to survive, (4) crossover and mutation, (5) the surviving and modified variable sets from the population. Through such operation, irrelevant spectral information is eliminated and the number of spectral variables is reduced.

3.1.2.5. Uninformative variable elimination (UVE)

The uninformative variable elimination (UVE) is a method for variable selection based on an analysis of regression coefficient of PLS. The UVE method was employed by Sun et al. Readers are referred to the corresponding references for details about many effective variable selection methods [52]. The method builds a large number of models with randomly selected calibration samples at first, and then each variable is evaluated with a stability of the corresponding coefficients in these models. Variables with poor stability are known as uninformative variable and are eliminated [53].

3.1.3. Calibration models

Multivariate regression techniques (quantitative analysis) aim to establish a relationship between the observed response values and spectral matrix. In our research, partial least squares (PLS) regression is a common multivariate method used in calibration of spectroscopy data. The principle of PLS is to use a linear least squares fitting technique. It builds linear models between an independent matrix X (spectral data) and a dependent matrix Y and estimate the regression coefficient matrix using least squares fitting techniques. Least squares support vector machines (LS-SVM) can deal with nonlinear relationships between variables.

3.1.3.1. Partial least squares (PLS)

Partial least squares (PLS) analysis is widely used for calibration in present chemometric analysis. It is an unsupervised statistical method used when not only a data array coming from X data is available but also a Y array that we want to predict from our X data [32]. Normally, there are two variable selection methods using PLS regression: using variable importance in projection scores and using regression coefficients estimated by PLS regression [54, 55]. The aim of PLS analysis is to find a latent variables linear regression model by projecting the X variables and the *Y* variables into a new latent space, where the covariance between these latent variables is maximized [1]. PLS analysis can be performed to establish the regression model leading to the content prediction of chemical components. PLS considers simultaneously the variable matrix *Y* (the values of SSC, pH) and the variable matrix *X* (the spectral data). Generally, the first step in PLS is to decompose the matrix, and the model is given:

$$X = TP + E \tag{2}$$

$$Y = UQ + F \tag{3}$$

In these equations, X is a $n \times m$ spectral matrix (n is the number of samples, m is the number of wavelengths), T and U are the score matrices of X matrix and Y matrix, P is the $m \times k$ matrix of X matrix and Q is the loading ($l \times k$), and y is the reference data ($n \times l$) that needs to be predicted from X (k is the number of latent variables), and E and F are the errors which come from the process of PLS regression [43]. The second step is that T and U are processed by linear regression. It must build the following linear correlation:

$$U = BT + E \tag{4}$$

where B represents the internal relations between U and T. In order to reach this object this object, the coordinate of T is rotated.

3.1.3.2. Least square support vector machine (LS-SVM)

Least square support vector machine (LS-SVM) is a set of related supervised learning method that analyzes data and recognizes patterns, and is used for classification and regression analysis. PLS method can only handle linear problems and builds a linear relationship between spectral variables and target chemical response such as SSC value. However, some researchers reported that the latent nonlinear information might be existed in the spectra data of fruit and the nonlinear models were better than linear models. The computational complexity and quality of the SVM does not directly depend on the dimension of input data. Therefore, least square support vector machine (LS-SVM) was applied to build a nonlinear model for a comparison of the prediction performance with linear PLS models. LS-SVM is widely applied in pattern recognition and function regression for the advantage of limited over-fitting, high predictive reliability and strong generalization ability [24]. More details of LS-SVM method can be found in the paper [56, 57]. The final LS-SVM regression model can be expressed as:

$$y(x) = \sum_{k=1}^{N} a_k K(x, x_k) + b \tag{5}$$

where $K(x, x_k)$ is the kernel function, x_k is the input vector, α_k is the Lagrange multiplier called support value, and b is the bias. The radial basis function (RBF), which is a frequently used kernel function $K(x, x_k)$, is used in this study and defined as follows:

$$K(x, x_k) = exp\left(-\frac{\|x_k - x\|^2}{2\sigma^2}\right) \tag{6}$$

In the equation, $\|x_k - x\|$ represents the distance between input vector and threshold vector, and σ is a width vector. Generally, the selected variables by wavelength selection methods could be used as the inputs to build the LS-SVM models.

3.1.4. Model validation

Validation procedures are crucial to assess the accuracy of the calibration and to avoid overfitting. The prediction ability of a calibration model can be evaluated by the correlation coefficient (*r*), root mean square error of prediction (*RMSEP*) and calibration (*RMSEC*) between the predicted value and the measured value in validation set [24]. In order to establish useful calibration models, different methods in spectral preprocessing and calibration modeling as mentioned above should be investigated. When cross validation is employed, the prediction performance could also be assessed by the root mean square error for cross validation (*RMSECV*). These indices are defined as follows:

$$r = \sqrt{\frac{\sum_{i=1}^{n}(\hat{y}_i - y_i)^2}{\sum_{i=1}^{n}(\hat{y}_i - y_{mean})^2}} \tag{7}$$

$$RMSEC = \sqrt{\frac{1}{n_c}\sum_{i=1}^{n_c}(\hat{y}_i - y_i)^2} \tag{8}$$

$$RMSEP = \sqrt{\frac{1}{n_p}\sum_{i=1}^{n_p}(\hat{y}_i - y_i)^2} \tag{9}$$

$$\text{bias} = \frac{1}{n}\sum_{i=1}^{n}(\hat{y}_i - y_i) \tag{10}$$

where, $\hat{y}_i$ is the predicted value of the *i*th observation, y_i is the measured value of the *i*th observation, y_m is the mean value of the calibration or prediction set, n, n_c, and n_p are respectively the number of observations in the data set, calibration and prediction set. Generally, a good model should have higher correlation coefficients, lower *RMSEC, RMSEP*, and *bias* values [58, 59].

3.2. Image processing and analysis techniques

Image processing and image analysis are considered to be the core of the hyperspectral imaging system with various algorithms and methods available to complete the specific classification and measurement. As illustrated in **Figure 9**, image processing and analysis are performed in three levels. The low level processing is the basic processing of image, which involves image acquisition and image preprocessing, and is the important step in image processing and analysis, which involves image segmentation, feature extraction, representation, and description [60]; the high level processing is the key step of image analysis, which involves recognition, interpretation and classification [2].

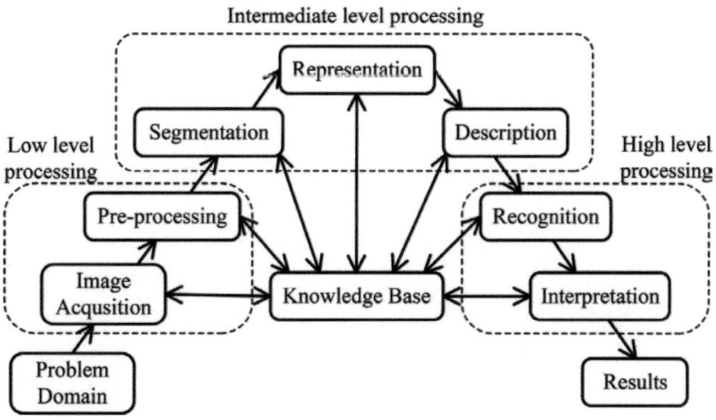

Figure 9. Different levels of image processing.

3.2.1. Image processing methods

The assessment accuracy of fruits and vegetables quality is highly related to the images. However, owing to the imperfections of the image acquisition systems, the images acquired are subject to various defects that will need subsequent processing. Image processing plays an important role in hyperspectral data analysis. The image processing involves a series of steps, which can be divided into three major steps: image preprocessing, segmentation and feature extraction [61].

The purpose of image preprocessing and calibration is to improve the quality of the obtained images by removing the noise, increasing the contrast and correcting the distortion [2]. Generally, the frequently used preprocessing methods include basic point operations (intensity mappings) and histogram equalization [43]. Basic point operations, such as luminance inversion and multiplicative brightness scaling, can improve by stretching the brightness levels into a mapping between the input level and the output level. Histogram equalization provides a sophisticated method for modifying the dynamic range and contrast of an image by changing the image so that its intensity histogram has a desired shape. Histogram model use nonlinear and nonmonotonic transfer functions to map the pixel intensity values of input and output images. Other typical image preprocessing techniques include filtering, transformation and arithmetic operations.

Image segmentation is the most vital and challenging step to partition the image into regions of interest (ROI). The goal of image segmentation is aimed at simplifying and altering the representation of an image into something more meaningful and easier to analyze. Image segmentation is typically used to locate objects and boundaries (lines, curves, etc.) in images. The accuracy of image segmentation plays an important role in the subsequent image processing. Threshold-based segmentation, edge-based segmentation, region-based segmentation, and classification-based segmentation are four major types of segmentation methods [62–64].

Feature extraction is a key step in connecting image processing and analysis, which converts image data or segmented regions into a set of feature vectors. In image processing, feature extraction builds features intended to be informative and nonredundant, facilitating the subsequent learning and generalization step [65]. When the image segmentation is successfully performed, if the data in ROI to an algorithm is too large to be processed, it can reduce its dimensionality. Feature extraction is related to dimensionality reduction. Thus, feature extraction is crucial to the accuracy of quality assessment. In general, shape features, texture features, color features and size features of the target are typically extracted for quality assessment.

3.2.2. Image analysis methods

Image analysis is a nondestructive method of calculating measurements and statistics based on the interesting values of images' pixels, and their corresponding spatial location. The image analysis is performed on the feature extracted from the image, and interprets the results. It uses intuitive explanations to display images and mathematically processing images, helping to solve the problem of computer vision. Vision measurement and pattern classification are the most crucial parts of image analysis.

Vision measurement is a quantitative analysis method in the image analysis. Visual measurement is the process of quantifying the parameters of interest from the features extracted from the image. It is the process of quantitative measurement of interest parameters based on the characteristics extracted from the image [66]. The computer vision systems can achieve different types of measurements. Generally, typical measurements include the size, texture and color.

Pattern classification, also known as pattern recognition, is a method for qualitative analysis in the image analysis. It is the science of reasoning based on measurement characteristics through probabilistic, statistical, computational geometry, multivariate analysis and algorithm design techniques. The classification techniques can be divided into two types: supervised methods and unsupervised methods. In the image analysis, the supervised methods are the most widely used. In most cases, the supervised classification method aims to build a model or a classifier for classification of labels according to the corresponding characteristics, while the unsupervised classification method is mainly used to classify image by finding out similarity between the selected features and using clustering algorithm. The widely used pattern classification methods in image analysis include Artificial Neural Network (ANN), Support vector machine (SVM), K-Nearest Neighbor (KNN), Adaptive Boosting, and decision tree. ANN is a nonlinear statistical data modeling tool that attempts to mimic the fault tolerance and capacity to learn biological neural systems by modeling the low-level structure of the brain [1]. ANN is widely used in hyperspectral image analysis, because it can handle a large amount of heterogeneous data with considerable flexibility and nonlinearity. It is composed of a set of interconnected artificial neurons, which are like a parallel system that capable of resolving the paradigm that linear computing cannot. SVM is a supervised nonparametric statistical learning model with associated learning algorithms that analyze data and recognize patterns, used for classification and regression analysis. In addition to performing linear classification, SVM can use the so-called kernel technique to efficiently perform nonlinear classification and map its inputs implicitly into high-dimensional feature spaces. As SVM, AdaBoost is one of the most successful supervised classification methods with the aim to maximize the minimum margin of a training sample [2]. KNN is another unsupervised classification method which is able to predict the response of the new sample by analyzing a certain number of the nearest neighbors in the feature space of the sample. In KNN, dataset is classified by minimizing the sum of squares of distances between each category and the corresponding cluster centroid [67]. Decision trees are commonly used in hyperspectral image analysis, to help identify a strategy that is most likely to reach a goal.

4. Applications in the quality assessment of fruits and vegetables

4.1. Applications of surface defect detection

The presence of surface defects influences the quality and price of fruits and vegetables, and weeding out the fruits and vegetables with serious defects early can prevent the infection of the whole patch. Therefore, detection of surface defects is the most commonly extended application of image and spectral analysis to the external quality inspection of fruits and vegetables.

Visual inspection of fruits and vegetables with respect to color, texture, size, and shape by traditional computer vision is already automated in the commercial sorting machines. However, sorting by defects is still a challenging task due to the high variance of defect types and existence of stem/calyx concavities [68]. The color, texture, or internal components of defects may be different from that of the sound; therefore, color, texture, or spectral reflectance are usually selected as the defect features to discriminate the defects from the sound peel. Many applications aimed to detect defects based on these features have been described by using hyperspectral or multispectral imaging system.

Due to lack of spectral information in conventional color images, traditional computer vision system is not efficient for the inspection of some defects with similar color and texture as sound peel, such as bruises, rottenness, or chilling injury. Hyperspectral and multispectral imaging systems provide powerful tools not only to detect skin defects but also to differentiate between a variety of defects that have similar color and texture or even to detect some defects that are not clearly visible [1]. Bruising is one of the familiar defects occurring on fruits and vegetables during post-harvest handling and processing stage. The existing commercial sorting machines are still not available in detecting bruises [69, 70]. An experiment of using a hyperspectral imaging system for bruise detection on apples was conducted by Xing et al. [70]. PCA and PLSDA were used to extract the spectral and spatial features from the hyperspectral images in the region between 400 and 1000 nm. Their experiment proved that combination of image processing and chemometric tools had a potential in detecting the bruises on apples. In order to detect the early bruises in apples, Baranowski et al. [69] proposed a system that included hyperspectral cameras equipped with sensors working in the visible and near-infrared (400–1000 nm), short-wavelength infrared (1000–2500 nm) and thermal imaging camera in mid-wavelength infrared (3500–5000 nm) ranges. Results showed that the principal components analysis (PCA) and minimum noise fraction (MNF) analyses of the images could make it possible to distinguish between areas with defects in the tissue and the sound ones, and the fast Fourier analysis of the image sequences after pulse heating of the fruit surface could provide additional information not only about the position of the area of damaged tissue but also about its depth. As unsupervised methods, the class number and the color or intensity value for each class are always randomly assigned by PCA and MNF. The robustness and stability of their algorithms are needed to be tested in inline inspection situation.

Decay is another common defect with great potential risk for consumers, sellers and growers. To fast detect and visualize the early decay in citrus, Li et al. [71] developed multispectral image processing method with mean normalization reducing spectral variability due to spherical fruit. The overall accuracy of 98.6% for test set with no false negatives was achieved. Their idea behind the proposed algorithm can be extended to detect the nonvisible damages of other fruit. Gómez-Sanchis et al. [72] presents the development of a hyperspectral system based on two liquid crystal tuneable filters for the acquisition of images of spherical fruits. They also designed a system that allows the filters to be exchanged quickly and without altering the acquired scene. The system and decay segmentation results are shown in **Figure 10**. Correctly classifying 98% of pixels as rotten or nonrotten tissues were achieved; however, changing the filters frequently decreases the detection efficiency, especially when working in the sorting line, the rotating products might cause the acquired scene vary with each of the filters.

Chilling injury is a common defect occurring during the storage and transportation at low temperatures. Liu et al. [73] developed a hyperspectral imaging system to detect the chilling injury in cucumber by using band ratio and PCA methods. Results revealed that either band ratio algorithm (Q811/756) or PCA transform in a spectral region between 733 and 848 nm could detect the chilling injury with an accuracy of over 90%. Ariana and Lu [74] found that the hyperspectral imaging under transmittance mode has shown potential for detecting internal

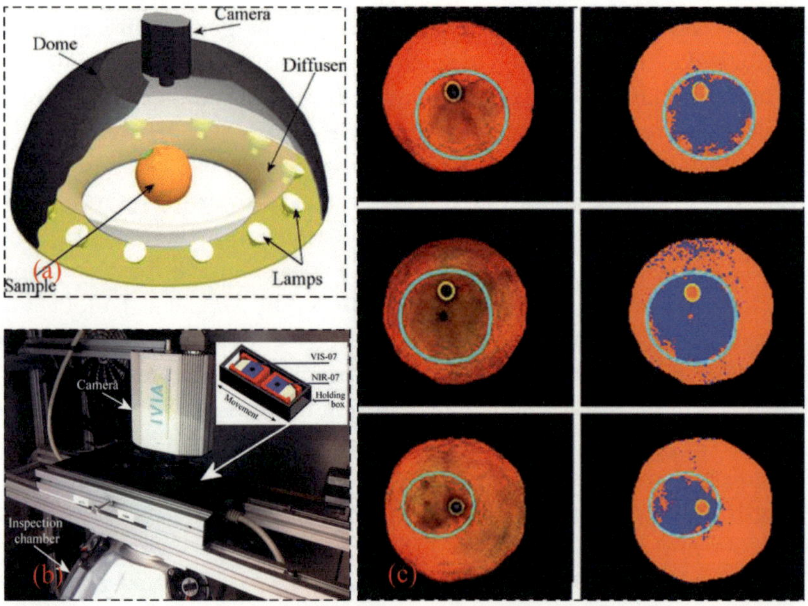

Figure 10. The system and decay segmentation results proposed by Gómez-Sanchis et al. (a) Hemispherical illumination chamber used to illuminate spherical objects (b) System created to facilitate the swap of two LCTF filters (c) RGB images and segmented images of mandarins with decay lesions [72].

defect. However, the technique still cannot meet the online speed requirement because of the need to acquire and analyze a large amount of image data. They determined up to four-waveband subsets by a branch and bound algorithm combined with the k-nearest neighbor classifier. The highest classification accuracies of 94.7 and 82.9% were achieved for cucumbers and whole pickles, respectively.

However, the acquisition and processing of the hyperspectral images is time-consuming, and the redundancy data makes the hyperspectral imaging system impossible to be used in-line or real-time. Actually, the hyperspectral imaging is always used for analysis and determining the effective wavelengths for a multispectral imaging system. Based on hyperspectral images and PCA, four efficient wavelengths (558, 678, 728, and 892 nm) were selected, and then a multispectral imaging system was developed by Xing et al. [75] to detect the bruises on apples. An overall accuracy of about 86% was obtained with their systems and algorithms. A near commercial multispectral imaging prototype for inline bruise detection was developed by Huang et al. [76] in NERCITA, China. Segmented principal component analysis (PCA) was conducted to eliminate data redundancy and select optimal wavelengths. Two dichroic beam-splitters, two band-pass filters with the center at selected wavelengths and three prism-based 2CCD multispectral progressive area scan cameras were used to develop the multispectral imaging system. Static and online tests were evaluated by their system, and 91.5% and 74.6% overall accuracy were achieved for static and online detection, respectively.

Table 1 shows a detailed summary of studies about the defect detection of fruits and vegetables by using hyperspectral imaging systems.

4.2. Applications of internal quality parameter measurement

4.2.1. Soluble solids content (SSC)

Soluble solids content, also named total soluble solids (TSS) content, is a collective index for sweetness measurement [77]. In the preharvest period, SSC profoundly dominates the optimal harvest time for various fruits and vegetables, whereas changes of SSC during the shelf-life period after harvesting would lead to quality fluctuation of fruits and vegetables [77]. Therefore, soluble solids content is an important internal quality attribute in determining fruit maturity and harvest time, and in assessing and grading post-harvest quality of apples [78].

In the past 20 years, many studies have been reported on predicting SSC in fruits using near-infrared spectroscopic technique. Leivavalenzuela et al. [79] made a report on the application of hyperspectral imaging technique for predicting the SSC of blueberries in the visible and short-wave near-infrared region of 500–1000 nm. In this study, Calibration models using partial least squares method were developed to predict the SSC, and the effect of fruit orientation on the model performance was evaluated. Results showed that hyperspectral imaging is promising for online sorting and grading of blueberries for firmness and perhaps SSC as well. Mendoza et al. [80] developed two different hyperspectral imaging systems: a stationary hyperspectral imaging system and a prototype on-line hyperspectral imaging system to evaluate the SSC in apples. The work used several methods, including discrete and continuous wavelet transform and conventional image texture analysis. Finally, the results showed

Products	Species	Applications	Types of CVS	Methods	Accuracy	Reference
Fruits	Apple	Quality grading	MIS	Statistical and syntactical classifiers	93.5%	[81]
	Apple	Defect segmentation	MIS	ANN	–	[68]
	Apple	Quality evaluation	MIS	Flat-field correction	95%	[82]
	Apple	Bruise detection	HIS	PCA, MNF, SIMCA, LDA, SVM	–	[69]
	Apple	Defect detection	HIS	ASD	–	[83]
	Apple	Defect and feces detection	MIS	_	–	[84]
	Apple	Bruise detection	HIS	PLS, SDA	93.95%	[85]
	Apple	Efficient wavelength selection	MIS	QDA	–	[86]
	Apple	Rottenness detection	HIS	LDA, CART	91.2%	[87]
	Apple	Bitter pit detection	HIS	PLS	–	[88]
	Apple	Defect detection	MIS	ANN	95.4%	[89]
	Apple	Defect detection	HIS	BR	99.5%	[90]
	Apple	Defect detection	HIS	SD, PCA	–	[83]
	Apple	Decayed spot, wound and rot detection	MIS	BR	92.42%	[91]
	Apple	Bruise detection	HIS	PCA, PLSDA	86%	[70]
	Apple	Bruise detection	MIS	PCA, MT	86%	[75]
	Apple	Bruise detection	HIS	PCA	>77.5%	[75]
	Apple	Bruise detection	HIS	PCA, MNT	88%, 94%	[92]
	Apple	Defect detection	MIS	Rotating	90%	[93]
	Apple	Defect detection	MIS	PCA, ANN	79%	[94]
	Apple	Chilling injury detection	HIS	ANN	98.4%	[95]
	Citrus	Canker detection	MIS	BR, T	95.3%	[96]
	Citrus	Skin damage detection	MIS	Bayesian discriminant analysis	86%	[97]
	Citrus	Common defect detection	HIS	PCA	93.7%	[98]
	Citrus	Light correction	HIS	Light correction	–	[87]
	Citrus	Rottenness detection	HIS	ANN, DT	98%	[99]

Products	Species	Applications	Types of CVS	Methods	Accuracy	Reference
	Pear	Bruise detection	HIS	PCA, MLC, EDC, MDC, SAM	93.8–95%	[100]
	Strawberry	Bruise detection	HIS	LDA, ND, ANN	100%	[101]
	Cherry	Pit detection	HIS	NN	97%	[102]
	Jujube	Insect infestation detection	HIS	JMP, MA	97%	[103]
Vegetables	Cucumber	Bruise detection	HIS	PCA, BR	75-95%	[89]
	Cucumber	Chilling injury detection	HIS	PCA, FLD	91%	[104]
	Cucumber	Chilling injury detection	HIS	BR, PCA	>90%	[75, 105]
	Mushroom	Bruise detection	HIS	PCA	79-100%	[106]
	Mushroom	Freeze damage detection	HIS	PCA, LDA	95%	[107]
	Mushroom	Enzymatic browning	HIS	PLS-DA	–	[108]
	Onion	Sour skin disease detection	HIS	MS	–	[109]

HIS: hyperspectral imaging system; MIS: multispectral imaging system; BR: band ratio; MS: mean reflectance spectra; ASD: asymmetric second difference; MT: moments thresholding; and T: thresholding.

Table 1. Summary of studies about the defect detection of fruits and vegetables.

that the integration of spectral and image features for hyperspectral scattering technique significantly improved firmness and SSC prediction (by the t-test) for all three cultivars but with a lesser degree of pronouncement for SSC.

Peng et al. [78] did a research on the hyperspectral imaging system for predicting soluble solids content (SSC) of "Golden Delicious" apples which was calibrated both spectrally and spatially. Their proposed methods, evaluating and comparing different mathematical models for describing the hyperspectral scattering profiles over the spectral region between 450 nm and 1000 nm coupled with the scattering profile correction methods, could improve the hyperspectral scattering technique for measuring fruit quality; and the study also showed the modified Lorentzian distribution function with three parameters without including the parameter for the asymptotic value which was most appropriate for predicting both fruit firmness and SSC. Rajkumar et al. [110] at three different temperatures used a hyperspectral imaging technique in the visible and NIR regions (400–1000 nm) to study bananas' SSC. Some quality parameters like moisture content were also determined and correlated with the spectral data using PLS. Their proposed methods, coupled with the scattering profile correction methods, could improve the hyperspectral scattering technique for measuring banana fruit quality.

Applications of hyperspectral imaging in fruit and vegetable SSC measurement could also be found in other crop products, such as strawberries, pears and so on [111, 112].

4.2.2. *Firmness*

Firmness is an important textural attribute for fruits and directly influences their shelf life and consumer acceptance, and it is an important internal quality attribute in determining fruit maturity and harvest time, and in assessing and grading post-harvest quality of apples. Thus, nondestructive sensing of fruit firmness would provide the fruit industry with a mean to ensure the quality and consistency of individual fruit, increase consumer satisfaction, and thus improve industry profitability [113].

Peng and Lu [113] proposed ten modified Lorentzian distribution with three parameters to characterize spatial scattering profiles from scattering images for Golden Delicious apples. A multilinear regression analysis was performed to predict the relationship between parameters of the scattering profile and the firmness of apples. This new method, coupled with the scattering profile correction methods, improved the hyperspectral scattering technique for measuring fruit and vegetable quality. Fan et al. [114] acquired hyperspectral reflectance image from each pear in visible and near-infrared (400–1000 nm) regions by employing the hyperspectral imaging system to determine SSC and firmness of pears. In this study, the variables selected by SPA, CARS and the combination of CARS and SPA were used for PLS regression. The overall results indicated that the CARS-SPA was an effective way for the selection of effective variables and the hyperspectral imaging system together with CARS-SPA-PLS model could be applied as a fast and potential method for the determination of SSC and firmness of pear. Qin et al. [115] measured the absorption and reduced scattering coefficients of apples through a spatially-resolved hyperspectral imaging technique and related them to fruit firmness. This research demonstrated the potential of using spectral absorption and scattering properties to evaluate internal quality attributes of horticultural products.

4.2.3. *Acidity/pH*

The acid content is often determined by a titratable method. A common method used for measuring ethylene production is to extract a gas sample from the internal core space of fruit or from a sealed container, in which fruits have been kept for a period of time and then analyzed using gas chromatograph [116]. The quality of fruit or vegetables is determined by a series of properties, such as acidity, which makes them attractive to consumers, is very crucial.

Cayuela et al. [117] described a portable AOTF-NIR spectrophotometer with a wide spectral range between 1100 and 2300 nm, which was equipped with a reflectance post-dispersive optical configuration and an InGaAs detector used for NIR prediction of fruit moisture content and free acidity. ElMasry et al. [111] determined acidity in strawberries by feat of a visible/NIR hyperspectral imaging system (400–1000 nm). It was found that the spectral pretreatments of mean-centering and automatic baseline correction enhanced PLS calibration model when compared with others pretreatments, such as Savitzky-Golay smoothing, MSC, and first and second derivatives.

Rungpichayapichet et al. [118] proposed a new hyperspectral imaging technology using a newly developed frame camera which was applied to determine internal properties of mango fruits including firmness, total soluble solids (TSS) and titratable acidity (TA). In their study, prediction models were developed using spectral data from relative surface reflectance of 160 fruits in the visible and NIR region of 450–998 nm analyzed by PLS regression. From their results, HSI can be used as a nondestructive technique for determining the quality of fruits which could potentially enhance grading capabilities in the industrial handling and processing of mango. Baiano et al. [119] carried out acidity determination in 7 cultivars of table grapes using NIR HSI with PLS models performing on the mean-centering correction spectra, and they achieved the coefficients of determination for predicting titratable acid and pH of red grapes and white grapes. They concluded that spectra information was not correlated with the sensory data, making hard prediction of attribute perception.

In addition to these fruits, the application of hyperspectral images acidity with broader range of 1000–2300 nm was acquired for the determination of total fat in beef cuts with good prediction abilities [120]. In other study, Abdel-Nour et al. [121] applied hyperspectral transmittance imaging (900–1700 nm) to classify eggs into three types with different docosahexaenoic acid contents using K-means analysis, resulting in 100% classification accuracy. Liu and Ngadi [122] detected fertility and early embryo development of chicken eggs using near-infrared hyperspectral imaging.

4.2.4. Moisture/water content

A fruit or vegetable consists of many different constituents, where water is the major component in fruits and vegetables [16]. Moisture content influences the taste, texture, weight, appearance, and shelf life of fruits and vegetables. Therefore, even a slight deviation from a defined standard can adversely impact the physical properties of a fruit or vegetable. For these reasons, the analysis to the moisture content of food products has a critical impact on quality and safety features [123].

Recently, hyperspectral imaging has also been used for determining the water content of other large variety of fruits and vegetables. Mollazade et al. [124] evaluated the potential of hyperspectral imaging combined with artificial neural networks to predict the moisture content in tomato fruit and to obtain spatial distribution maps. Their works displayed the spatial distribution of moisture content as a color map, where colors represent different values of predicted attribute. Finally, result showed that the feasibility of the method for characterizing the spatial distribution of an attribute in horticultural produce. Dong and Guo [125] proposed a hyperspectral reflectance imaging technology in near-infrared regions (900–1,70,002 nm) to evaluate soluble solids content (SSC), firmness, moisture content, and pH values of "Fuji" apples. They employed PLS regression, LS-SVM and back propagation (BP) network modeling methods to establish models to predict SSC, firmness, MC, and pH of apples, respectively. Results indicated that the moisture content could be predicted exactly by all developed models.

Firtha et al. [126] described an approach for the prediction of moisture content in carrot tissue. The work reduced the data load of hyperspectral experiments by using sample-specific vector-to-scalar operators for real-time feature extraction and a systematic procedure for compensating for pixels in the NIR sensor. Results demonstrated that the approach to predict the moisture content of carrots is feasible. Except what we mentioned above, hyperspectral imaging can be applied on the moisture content of all kinds of fruits and vegetables such as strawberries and soybean [24, 127].

4.2.5. Starch content

Starch is the main form of carbohydrate in our food, which is present in a variety of grains, vegetables and fruits. During the ripening of fruit, starch is changed into sugar, which gives sweetness to ripe fruits [128]. The harvest time of fruits, matching the desired commercial characteristics, is assessed through starch-iodine test in practice [129].

Peirs et al. [130] employed a threshold value of the first principal component score image to measure the starch distribution and starch index of apple fruit during maturation. Results showed that the starch concentration obtained in each position of the fruit was continuously measured compared with the discrete values obtained with the traditional technique. The method that they are proposed will speed up the application while the purchase costs decrease considerably and can be considered as a model system to map quality attributes of fruits. Menesatti et al. [129] researched the relationships of near-infrared (NIR) spectral images, starch/starch-free patterns visually assessed and RGB color images through PLS-DA to assess the starch index of apples. They studied the spectral region between 1000 and 1,70,002 nm through PLS-DA to assess the starch index in apples. Their proposed methods, avoiding expert's subjective interpretation of starch index assignment, show the feasibility of NIR imaging spectroscopy as a tool for fruit maturity determination.

Chen et al. [131] studied nondestructive detection of starch content in potatoes using the SPA-MLR model and SPA-PLSR model, respectively. Results showed that the effect of the SPA-MLR model was superior to that of the SPA-PLSR model. Trong et al. [132] employed the starch index to estimate the optimal cooking time of potatoes. The changes caused by the microstructure and composition of starch affected the interaction of light with the starch granules at different regions inside the potatoes. In their research, the potential of hyperspectral imaging in the wavelength range of 400 nm to 1000 nm in combination with chemometric tools and image processing for contactless detection of the cooking front in potatoes has been investigated.

4.2.6. Ripening/maturity stages

The definition of apple maturity corresponding to the stage of fruit development, giving minimum acceptable quality to the ultimate consumer, implies measurable points in the commodity's development and the need for techniques to measure maturity [133]. In addition, concerning the internal quality, maturity is extremely important to determine the harvest time and optimize the post-harvest treatment and environment [1, 16].

In recent years, many works on the determination of the maturity of fruits have been reported. An example of such studies is that of Rajkumar et al. [110] who studied banana fruit quality and maturity stages at three different temperatures by using hyperspectral imaging technique in the visible and near-infrared (400–1000 nm) regions to determine the quality parameters like moisture content. In their research, they concluded that the change in TSS and firmness of banana fruits stored at different temperatures during the ripening process followed the polynomial relationships and the change in moisture content followed a linear relationship at different maturity stages. And Garridonovell et al. [134] evaluated the potential of RGB digital imaging and hyperspectral imaging for discriminating maturity level in apples. In their research, segmentation, preprocessing and PLS-DA are applied to hyperspectral data analysis, while illumination correction, dimensionality reduction and linear discriminant analysis (LDA) are applied to RGB data analysis. Finally, they concluded that hyperspectral discrimination classified different storage regimes better than RGB.

Herrerolangreo et al. [135] developed an automatic procedure which is able to classify commercial peaches according to their maturity stage through multispectral imaging techniques. They proposed and validated the process of evaluating peach maturity through spectral imaging, which is very crucial for ensuring its quality of optimum peach ripeness. The proposed method is nondestructive and quick, and thus, it will have a good perspective for its application in fresh fruit packing lines. Girod et al. [136] introduced a nondestructive and quick technique that can measure the DM content to assess the maturity of avocados. The work analyzed avocado fruits at different maturity stages through hyperspectral imaging in reflectance and absorbance modes. The proposed method indicated that the reasonably accurate models could be obtained for DM content with the entire spectral range. Applications of hyperspectral imaging to measure maturity stages of fruit and vegetable could also be found in pawpaws, tomatoes and grapes [1, 137, 138].

5. Conclusions

Over the past decades, hyperspectral imaging technique has been rapidly developing and widely applied in nondestructive fruit and vegetable quality assessment. This chapter provides the principles, developments and applications of hyperspectral imaging technology in the quality inspection of fruits and vegetables. The principal components, basic theories and corresponding processing and analytical methods are also reported in this chapter. Looking into the future of fast inline sorting industry, hyperspectral imaging faces both challenges and opportunities. The challenges include the influence of physical and biological variability, whole surface detection, discrimination between defects and stems/calyxes, unobvious defect detection, robustness of the features and algorithms, as well as rapid multispectral imaging system development. Though many solutions have been presented to solve the challenging problems in fruit and vegetable quality inspection by using hyperspectral imaging technique in previous studies by the scientific researchers worldwide, the challenges presented above will continue to be intractable problems for a long time.

Author details

Xiaona Li, Ruolan Li, Mengyu Wang, Yaru Liu, Baohua Zhang* and Jun Zhou

*Address all correspondence to: bhzhang@njau.edu.cn

College of Engineering, Nanjing Agricultural University, Nanjing, Jiangsu, PR China

References

[1] Lorente D, Aleixos N, Gómez-Sanchis J, et al. Recent Advances and Applications of Hyperspectral Imaging for Fruit and Vegetable Quality Assessment. An Introduction to Quantum Computing Algorithms. Birkhauser; 2012. pp. 231-252

[2] Zhang B, Huang W, Li J, Zhao C, Fan S, Wu J, Liu C. Principles, developments and applications of computer vision for external quality inspection of fruits and vegetables: A review. Food Research International. 2014;**62**:326-343

[3] Costa C, Antonucci F, Pallottino F, Aguzzi J, Sun DW, Menesatti P. Shape analysis of agricultural products: A review of recent research advances and potential application to computer vision. Food & Bioprocess Technology. 2011;**4**:673-692

[4] Cubero S, Aleixos N, Moltó E, Gómez-Sanchis J, Blasco J. Advances in machine vision applications for automatic inspection and quality evaluation of fruits and vegetables. Food & Bioprocess Technology. 2011;**4**:487-504

[5] Goetz AF, Vane G, Solomon JE, Rock BN. Imaging spectrometry for Earth remote sensing. Science. 1985;**228**:1147-1153

[6] Arngren M, Schmidt MN, Larsen J. Unmixing of Hyperspectral images using Bayesian non-negative matrix factorization with volume prior. Journal of Signal Processing Systems. 2011;**65**:479-496

[7] Monteiro ST, Minekawa Y, Kosugi Y. Prediction of sweetness and amino acid content in soybean crops from hyperspectral imagery. Isprs Journal of Photogrammetry & Remote Sensing. 2007;**62**:2-12

[8] Smail VW, Fritz AK, Wetzel DL. Chemical imaging of intact seeds with NIR focal plane array assists plant breeding. Vibrational Spectroscopy. 2006;**42**:215-221

[9] Uno Y, Prasher SO, Lacroix R, Goel PK, Karimi Y, Viau A, Patel RM. Artificial neural networks to predict corn yield from compact airborne spectrographic imager data. Computers & Electronics in Agriculture. 2005;**47**:149-161

[10] Chang CI. Hyperspectral Imaging: Techniques for Spectral Detection and Classification. Plenum Publishing Co; 2003

[11] Qin J, Chao K, Kim MS, Lu R, Burks TF. Hyperspectral and multispectral imaging for evaluating food safety and quality. Journal of Food Engineering. 2013;**118**:157-171

[12] Zeng XA. Recent developments and applications of hyperspectral imaging for quality evaluation of agricultural products: A review. Critical Reviews in Food Science & Nutrition. 2015;**55**:1744

[13] Akodagali J, Balaji S. Computer vision and image analysis based techniques for automatic characterization of fruits a review. Biotechnology and Bioengineering. 2012;**38**:1001-1006

[14] Kathman A Optical device and associated methods. In: US. US8411379[P]. 2013

[15] Ko CH. Optical Wavelength Dispersion Device and Method of Manufacturing the Same. 2017

[16] Wu D, Sun DW. Advanced applications of hyperspectral imaging technology for food quality and safety analysis and assessment: A review – Part II: Applications. Innovative Food Science & Emerging Technologies. 2013;**19**:15-28

[17] Du CJ, Sun DW. Recent developments in the applications of image processing techniques for food quality evaluation. Trends in Food Science & Technology. 2004;**15**:230-249

[18] Chow RH, Hwang JY, Lee NS, Shung KK, Weitz AC. System and method for determining tumor invasiveness. US 20140087411 A1 [P]. 2014

[19] Liu D, Zeng XA, Sun DW. Recent developments and applications of hyperspectral imaging for quality evaluation of agricultural products: A review. Critical Reviews in Food Science & Nutrition. 2015;**55**:1744

[20] Liu Z, Jing W. Hyperspectral endmember detection method based on Bayesian Decision Theory. In: Software Engineering and Knowledge Engineering: Theory and Practice. Springer Berlin Heidelberg; 2012. pp. 727-732

[21] Elmasry G, Kamruzzaman M, Sun D, Allen P. Principles and applications of hyperspectral imaging in quality evaluation of agro-food products: A review. Critical Reviews in Food Science & Nutrition. 2012;**52**:999

[22] Patel YG, Rajadhyaksha M, Dimarzio CA. Optimization of pupil design for point-scanning and line-scanning confocal microscopy. Biomedical Optics Express. 2011;**2**: 2231

[23] Wilson T, Hewlett SJ. Imaging in scanning microscopes with slit-shaped detectors. Journal of Microscopy. 1990;**160**:115-139

[24] Wang H, Peng J, Xie C, Bao Y, Yong H. Fruit quality evaluation using spectroscopy technology: A review. Sensors. 2015;**15**:11889

[25] Huang W, Zhang B, Li J, et al. Early detection of bruises on apples using near-infrared hyperspectral image [C]. International Conference on Photonics and Image in Agriculture Engineering. 2013:87610P

[26] Lee WH, Kim MS, Lee H, Delwiche SR, Bae H, Kim DY, Cho BK. Hyperspectral near-infrared imaging for the detection of physical damages of pear. Journal of Food Engineering. 2014;**130**:1-7

[27] Fox G, Manley M. Applications of single kernel conventional andhyperspectral imaging near infrared spectroscopy in cereals. Journal of the Science of Food & Agriculture. 2014;**94**:174-179

[28] Zhang B, Fan S, Li J, Huang W, Zhao C, Qian M, Zheng L. Detection of early rottenness on apples by using hyperspectral imaging combined with spectral analysis and image processing. Food Analytical Methods. 2015;**8**:2075-2086

[29] Zhang X, Chen S, Ling Z, Zhou X, Ding DY, Kim YS, Xu F. Method for removing spectral contaminants to improve analysis of Raman imaging data. Scientific Reports. 2017;**7**:39891

[30] Magwaza LS, Opara UL, Nieuwoudt H, Cronje PJR, Saeys W, Nicolaï B. NIR spectroscopy applications for internal and external quality analysis of citrus fruit – A review. Food & Bioprocess Technology. 2012;**5**:425-444

[31] Rinnan Å, Berg FVD, Engelsen SB. Review of the most common pre-processing techniques for near-infrared spectra. TrAC – Trends in Analytical Chemistry. 2009;**28**:1201-1222

[32] Nicolaï BM, Beullens K, Bobelyn E, Peirs A, Saeys W, Theron KI, Lammertyn J. Nondestructive measurement of fruit and vegetable quality by means of NIR spectroscopy: A review. Postharvest Biology & Technology. 2007;**46**:99-118

[33] Kim JH, Jeung GW, Lee JW, Kim KS. Performance Evaluation of a Two-Dimensional Savitzky-Golay Filter for Image Smoothing Applications. 2016

[34] Sun T, Xu WL, Lin JL, Liu MH, He XW. Determination of soluble solids content in navel oranges by Vis/NIR diffuse transmission spectra combined with CARS method. Spectroscopy & Spectral Analysis. 2012;**32**:3229-3233

[35] Barnes R, Dhanoa M, Lister S. Letter: Correction to the description of standard normal variate (SNV) and de-trend (DT) ransformations in practical spectroscopy with applications in food and everage analysis – 2nd ed. Journal of Near Infrared Spectroscopy. 1993;**1**:185-186

[36] Dhanoa MS, Barnes RJ, Lister SJ. Standard normal variate transformation and de-trending of near-infrared diffuse reflectance spectra. Applied Spectroscopy. 1989;**43**:772-777

[37] Maleki MR, Mouazen AM, Ramon H, Jde B. Multiplicative scatter correction during on-line measurement with near infrared spectroscopy. Biosystems Engineering. 2007;**96**:427-433

[38] Chen JY, Zhang H, Ma J, Tuchiya T, Miao Y. Determination of the degree of degradation of frying rapeseed oil using Fourier-transform infrared spectroscopy combined with partial least-squares regression. International Journal of Analytical Chemistry. 2015;**2015**:185367

[39] Workman JJ, Springsteen AW. Applied spectroscopy: A compact reference for practitioners. 1998

[40] Ganesh A, Jena SK, Balasubramanian G, Pradhan N. A comparison study of function approximation using Fourier and Wavelet transforms. 2011:784-787

[41] Sun T, Lin H, Xu H, Ying Y. Effect of fruit moving speed on predicting soluble solids content of 'Cuiguan' pears (*Pomaceae pyrifolia* Nakai cv. Cuiguan) using PLS and LS-SVM regression. Postharvest Biology & Technology. 2009;**51**:86-90

[42] Hui L. Non-destructive detection of kiwifruit firmness based on near-infrared diffused spectroscopy. Transactions of the Chinese Society for Agricultural Machinery. 2011;**42**:145-149

[43] Liu D, Sun DW, Zeng XA. Recent advances in wavelength selection techniques for hyperspectral image processing in the food industry. Food & Bioprocess Technology. 2014;**7**:307-323

[44] Li H, Liang Y, Xu Q, Cao D. Key wavelengths screening using competitive adaptive reweighted sampling method for multivariate calibration. Analytica Chimica Acta. 2009;**648**:77

[45] Yun Y, Wei Y, Zhao X, Wu W, Liang Y, Lu H. A green method for the quantification of polysaccharides in *Dendrobium officinale*. RSC Advances. 2015;**5**:105057-105065

[46] Yang Y, Jin Y, Wu Y, Chen Y. (2016). Application of near infrared spectroscopy combined with competitive adaptive reweighted sampling partial least squares for on-line monitoring of the concentration process of Wangbi tablet 24

[47] Li HD, Xu QS, Liang YZ. Random frog: An efficient reversible jump Markov Chain Monte Carlo-like approach for variable selection with applications to gene selection and disease classification. Analytica Chimica Acta. 2012;**740**:20-26

[48] Yun YH, Li HD, Wood LRE, Fan W, Wang JJ, Cao DS, Xu QS, Liang YZ. An efficient method of wavelength interval selection based on random frog for multivariate spectral calibration. Spectrochimica Acta Part A Molecular & Biomolecular Spectroscopy. 2013;**111**:31

[49] Araújo MCU, Saldanha TCB, Galvão RKH, Yoneyama T, Chame HC, Visani V. The successive projections algorithm for variable selection in spectroscopic multicomponent analysis. Chemometrics and Intelligent Laboratory Systems. 2001;**57**:65-73

[50] Wu D, Sun DW, He Y. Application of long-wave near infrared hyperspectral imaging for measurement of color distribution in salmon fillet. Innovative Food Science & Emerging Technologies. 2012;**16**:361-372

[51] Mehmood T, Liland KH, Snipen L, Sæbø S. A review of variable selection methods in partial least squares regression. Chemometrics and Intelligent Laboratory Systems. 2012;**118**:62-69

[52] And VC, Massart D, Noord OED, And SDJ, Vandeginste BM, Sterna C. Elimination of uninformative variables for multivariate calibration. Analytical Chemistry. 1996;**68**:3851

[53] Cai W, Li Y, Shao X. A variable selection method based on uninformative variable elimination for multivariate calibration of near-infrared spectra. Chemometrics and Intelligent Laboratory Systems. 2008;**90**:188-194

[54] Giuseppe P, Paolo P, Hans-Dieter Z. Performance of PLS regression coefficients in selecting variables for each response of a multivariate PLS for omics-type data. Advances & Applications in Bioinformatics & Chemistry – AABC. 2009;**2**:57-70

[55] Mehmood T, Martens H, Sæbø S, Warringer J, Snipen L. A partial least squares based algorithm for parsimonious variable selection. Algorithms for Molecular Biology Amb. 2011;**6**:27

[56] Liu F, He Y, Wang L, Pan H. Feasibility of the use of visible and near infrared spectroscopy to assess soluble solids content and pH of rice wines. Journal of Food Engineering. 2007;**83**:430-435

[57] Suykens JAK, Vandewalle J. Least squares support vector machine classifiers. Neural Processing Letters. 1999;**9**:293-300

[58] Lin S, Huang X. Advances in computer science, environment, ecoinformatics, and education. In: International Conference, CSEE 2011, Wuhan, China, August 21-22, 2011. Proceedings, Part IV. Communications in Computer & Information Science. 2011. p. 218

[59] Shao Y, Yong H. Visible/near infrared spectroscopy and chemometrics for the prediction of trace element (Fe and Zn) levels in rice leaf. Sensors. 2013;**13**:1872

[60] Zou X, Zhao J. Nondestructive Measurement in Food and Agro-Products. 2015

[61] Sharma N, Ray AK, Sharma S, Shukla KK, Pradhan S, Aggarwal LM. Segmentation and classification of medical images using texture-primitive features: Application of BAM-type artificial neural network. Journal of Medical Physics. 2008;**33**:119-126

[62] Jackman P, Sun DW, Allen P. Recent advances in the use of computer vision technology in the quality assessment of fresh meats. Trends in Food Science & Technology. 2011;**22**:185-197

[63] Narendra VG, Hareesh KS. Quality inspection and grading of agricultural and food products by computer vision-a review. International Journal of Computer Applications. 2010;**2**:43-65

[64] Teena M, Manickavasagan A, Mothershaw A, Hadi SE, Jayas DS. Potential of machine vision techniques for detecting Fecal and microbial contamination of food products: A review. Food & Bioprocess Technology. 2013;**6**:1621-1634

[65] Kamila NK. Handbook of Research on Emerging Perspectives in Intelligent Pattern Recognition, Analysis, and Image Processing: Information Science Reference Imprint of: IGI Publishing. 2016

[66] Park B, Lu R. Hyperspectral imaging technology in food and agriculture. Food Engineering. 2015

[67] Shawe-Taylor J, Cristianini N. Kernel methods for pattern analysis. Journal of the American Statistical Association. 2004;**101**:1730-1730

[68] Unay D, Gosselin B. Automatic defect segmentation of 'Jonagold' apples on multi-spectral images: A comparative study. Postharvest Biology & Technology. 2006;**42**:271-279

[69] Baranowski P, Mazurek W, Wozniak J, Majewska U. Detection of early bruises in apples using hyperspectral data and thermal imaging. Journal of Food Engineering. 2012;**110**:345-355

[70] Xing J, Saeys W, Baerdemaeker JD. Combination of chemometric tools and image processing for bruise detection on apples. Computers & Electronics in Agriculture. 2007;**56**:1-13

[71] Li J, Tian X, Huang W, Zhang B, Fan S. Application of long-wave near infrared hyperspectral imaging for measurement of soluble solid content (SSC) in pear. Food Analytical Methods. 2016;**9**:3087-3098

[72] Gómez-Sanchis J, Lorente D, Soria-Olivas E, Aleixos N, Cubero S, Blasco J. Development of a hyperspectral computer vision system based on two liquid crystal tuneable filters for fruit inspection. Application to detect citrus fruits decay. Food & Bioprocess Technology. 2014;**7**:1047-1056

[73] Liu Y, Chen YR, Wang CY, Chan DE, Kim MS. Development of hyperspectral imaging technique for the detection of chilling injury in cucumbers; spectral and image analysis. Applied Engineering in Agriculture. 2006;**22**:101-111

[74] Ariana DP, Lu RF. Evaluation of internal defect and surface color of whole pickles using hyperspectral imaging. Journal of Food Engineering. 2010;**96**:583-590

[75] Xing J, Bravo C, Jancsok PT, Ramon H, De Baerdemaeker J. Detecting bruises on 'Golden Delicious' apples using hyperspectral imaging with multiple wavebands. Biosystems Engineering. 2005;**90**:27-36

[76] Huang Q, Chen Q, Li H, et al. Non-destructively sensing pork's freshness indicator using near infrared multispectral imaging technique. Rsc Advances. 2015;**5**:95903-95910

[77] Pu YY, Sun DW, Riccioli C, et al. Calibration transfer from micro NIR spectrometer to hyperspectral imaging: A case study on predicting soluble solids content of bananito fruit (Musa acuminata). Food Analytical Methods. 2017;1-13

[78] Peng Y, Lu R. Analysis of spatially resolved hyperspectral scattering images for assessing apple fruit firmness and soluble solids content. Postharvest Biology & Technology. 2008;**48**:52-62

[79] Leivavalenzuela GA, Lu R, Aguilera JM. Prediction of firmness and soluble solids content of blueberries using hyperspectral reflectance imaging. Journal of Food Engineering. 2013;**115**:91-98

[80] Mendoza F, Lu R, Ariana D, et al. Integrated spectral and image analysis of hyperspectral scattering data for prediction of apple fruit firmness and soluble solids content. Postharvest Biology & Technology. 2011;**62**:149-160

[81] Unay D, Gosselin B, Kleynen O, Leemans V, Destain MF, Debeir O. Automatic grading of bi-colored apples by multispectral machine vision. Computers and Electronics in Agriculture. 2011;**75**:204-212

[82] Throop JA, Aneshansley DJ, Anger WC, Peterson DL. Quality evaluation of apples based on surface defects: Development of an automated inspection system. Postharvest Biology and Technology. 2005;**36**:281-290

[83] Mehl PM, Chen YR, Kim MS, Chan DE. Development of hyperspectral imaging technique for the detection of apple surface defects and contaminations. Journal of Food Engineering. 2004;**61**:67-81

[84] Kim MS, Cho BK, Lefcourt AM, Chen YR, Kang S. Multispectral fluorescence lifetime imaging of feces-contaminated apples by time-resolved laser-induced fluorescence imaging system with tunable excitation wavelengths. Applied Optics. 2008;**47**:1608-1616

[85] Elmasry G, Wang N, Vigneault C, et al. Early detection of apple bruises on different background colors using hyperspectral imaging. LWT – Food Science and Technology. 2008;**41**:337-345

[86] Kleynen O, Leemans V, Destain MF. Selection of the most efficient wavelength bands for 'Jonagold' apple sorting. Postharvest Biology and Technology. 2003;**30**:221-232

[87] Gomez-Sanchis J, Gomez-Chova L, Aleixos N, Camps-Valls G, Montesinos-Herrero C, Molto E, Blasco J. Hyperspectral system for early detection of rottenness caused by *Penicillium digitatum* in mandarins. Journal of Food Engineering. 2008;**89**:80-86

[88] Nicolaï BM, Lotze E, Peirs A Scheerlinck N, Theron KI. Non-destructive measurement of bitter pit in apple fruit using NIR hyperspectral imaging. Postharvest Biology and Technology. 2006;**40**:1-6

[89] Ariana D, Guyer DE, Shrestha B. Integrating multispectral reflectance and fluorescence imaging for defect detection on apples. Computers and Electronics in Agriculture. 2006;**50**:148-161

[90] Kim MS, Chen Y-R, Cho B-K, Chao K, Yang C-C, Lefcourt AM, Chan D. Hyperspectral reflectance and fluorescence line-scan imaging for online defect and fecal contamination inspection of apples. Sensing and Instrumentation for Food Quality and Safety. 2007;**1**:151-159

[91] Lee D-J, Schoenberger R, Archibald J, McCollum S. Development of a machine vision system for automatic date grading using digital reflective near-infrared imaging. Journal of Food Engineering. 2008;**86**:388-398

[92] Lu R. Detection of bruises on apples using near-infrared hyperspectral imaging. Transactions of the ASAE. 2003;**46**:523-530

[93] Bennedsen BS, Peterson DL, Tabb A. Identifying defects in images of rotating apples. Computers and Electronics in Agriculture. 2005;**48**:92-102

[94] Bennedsen BS, Peterson DL, Tabb A. Identifying apple surface defects using principal components analysis and artificial neural networks. Transactions of the ASABE. 2007;**50**:2257-2265

[95] ElMasry G, Wang N, Vigneault C. Detecting chilling injury in red delicious apple using hyperspectral imaging and neural networks. Postharvest Biology and Technology. 2009;**52**:1-8

[96] Qin JW, Burks TF, Zhao XH, Niphadkar N, Ritenour MA. Development of a two-band spectral imaging system for real-time citrus canker detection. Journal of Food Engineering. 2012;**108**:87-93

[97] Blasco J, Cubero S, Gomez-Sanchis J, Mira P, Molto E. Development of a machine for the automatic sorting of pomegranate (Punica Granatum) arils based on computer vision. Journal of Food Engineering. 2009;**90**:27-34

[98] Li JB, Rao XQ, Ying YB. Detection of common defects on oranges using hyperspectral reflectance imaging. Computers and Electronics in Agriculture. 2011;**78**:38-48

[99] Gomez-Sanchis J, Martin-Guerrero JD, Soria-Olivas E, Martinez-Sober M, Magdalena-Benedito R, Blasco J. Detecting rottenness caused by Penicillium genus fungi in citrus fruits using machine learning techniques. Expert Systems with Applications. 2012;**39**:780-785

[100] Zhao JW, Ouyang Q, Chen QS, Wang JH. Detection of bruise on pear by hyperspectral imaging sensor with different classification algorithms. Sensor Letters. 2010;**8**:570-576

[101] Nagata M, Tallada JG, Kobayashi T. Bruise detection using NIR hyperspectral imaging for strawberry (Fragaria × ananassa Duch.). Environment Control in Biology. 2006;**44**:133

[102] Qin J, Lu R. Detection of pits in tart cherries by hyperspectral transmission imaging. Transactions of the ASAE. 2005;**48**:1963-1970

[103] Wang J, Nakano K, Ohashi S, et al. Detection of external insect infestations in jujube fruit using hyperspectral reflectance imaging. Biosystems Engineering. 2011;**108**:345-351

[104] Cheng X, Chen YR, Tao Y, Wang CY, Kim MS, Lefcourt AM. A novel integrated PCA and FLD method on hyperspectral image feature extraction for cucumber chllling damage inspection. Transactions of the ASAE. 2004;**47**:1313-1320

[105] Liu YL, Chen YR, Wang CY, Chan DE, Kim MS. Development of a simple algorithm for the detection of chilling injury in cucumbers from visible/near-infrared hyperspectral imaging. Applied Spectroscopy. 2005;**59**:78-85

[106] Gowen AA, O'Donnell CP, Taghizadeh M, Cullen PJ, Frias JM, Downey G. Hyperspectral imaging combined with principal component analysis for bruise damage detection on white mushrooms (*Agaricus bisporus*). Journal of Chemometrics. 2008;**22**:259-267

[107] Gowen AA, Taghizadeh M, O'Donnell CP. Identification of mushrooms subjected to freeze damage using hyperspectral imaging. Journal of Food Engineering. 2009;**93**:7-12

[108] Taghizadeh M, Gowen AA, O'Donnell CP. The potential of visible-near infrared hyperspectral imaging to discriminate between casing soil, enzymatic browning and undamaged tissue on mushroom (*Agaricus bisporus*) surfaces. Computers and Electronics in Agriculture. 2011;**77**:74-80

[109] Wang Y, Zhang M, Mujumdar AS. Influence of green banana flour substitution for cassava starch on the nutrition, color, texture and sensory quality in two types of snacks. LWT-Food Science and Technology. 2012;**47**:175-182

[110] Rajkumar P, Wang N, Eimasry G, Gsv R, Gariepy Y. Studies on banana fruit quality and maturity stages using hyperspectral imaging. Journal of Food Engineering. 2012;**108**:194-200

[111] Elmasry G, Wang N, Elsayed A, et al. Hyperspectral imaging for nondestructive determination of some quality attributes for strawberry. Journal of Food Engineering. 2007;**81**:98-107

[112] Li J, Zhang B, Zhao C, et al. Qualitative analysis of soluble solid content and firmness of pear based on successive projections algorithm and least square support vector machine. Sensor Letters. 2014;**12**:575-580 (576)

[113] Peng Y, Lu R. Improving apple fruit firmness predictions by effective correction of multispectral scattering images. Postharvest Biology & Technology. 2006;**41**:266-274

[114] Fan S, Huang W, Guo Z, Zhang B, Zhao C. Prediction of soluble solids content and firmness of pears using hyperspectral reflectance imaging. Food Analytical Methods. 2015;**8**:1936-1946

[115] Qin J, Lu R, Peng Y. Internal quality evaluation of apples using spectral absorption and scattering properties. Proceedings of SPIE. 2007;**6761**, 67610M-67610M-67611

[116] Noh HK, Lu R. Hyperspectral laser-induced fluorescence imaging for assessing apple fruit quality. Postharvest Biology & Technology. 2007;**43**:193-201

[117] Cayuela JA, Garc AJM, Caliani N. NIR prediction of fruit moisture, free acidity and oil content in intact olives. Grasas Y Aceites. 2009;**60**:194-202

[118] Rungpichayapichet P, Nagle M, Yuwanbun P, Khuwijitjaru P, Mahayothee B, Müller J. Prediction mapping of physicochemical properties in mango by hyperspectral imaging. Biosystems Engineering. 2017;**159**:109-120

[119] Baiano A, Terracone C, Peri G, et al. Application of hyperspectral imaging for prediction of physico-chemical and sensory characteristics of table grapes. Computers & Electronics in Agriculture. 2012;**87**:142-151

[120] Kobayashi K, Matsui Y, Maebuchi Y, et al. Near infrared spectroscopy and hyperspectral imaging for prediction and visualisation of fat and fatty acid content in intact raw beef cuts. Journal of Near Infrared Spectroscopy. 2010;**18**:301-315

[121] Abdel-Nour N, Ngadi M. Detection of omega-3 fatty acid in designer eggs using hyperspectral imaging. International Journal of Food Sciences & Nutrition. 2011;**62**:418-422

[122] Liu L, Ngadi MO. Detecting fertility and early embryo development of chicken eggs using near-infrared hyperspectral imaging. Food & Bioprocess Technology. 2013;**6**:2503-2513

[123] Pu YY, Feng YZ, Sun DW. Recent progress of hyperspectral imaging on quality and safety inspection of fruits and vegetables: A review. Comprehensive Reviews in Food Science & Food Safety. 2015;**14**:176-188

[124] Mollazade K, Omid M, Akhlaghian-Tab F, Mohtasebi SS, Zude M. Spatial mapping of moisture content in tomato fruits using hyperspectral imaging and artificial neural networks. In: CIGR-Ageng2012: IV International workshop on Computer Image Analysis in Agriculture. 2012

[125] Dong J, Guo W, Wang Z, et al. Nondestructive determination of soluble solids content of 'Fuji' apples produced in different areas and bagged with different materials during ripening. Food Analytical Methods. 2016;**9**:1087-1095

[126] Firtha F, Fekete A, Kaszab T, Gillay B, Nogulanagy M, Kovács Z, Kantor DB. Methods for improving image quality and reducing data load of NIR hyperspectral images. Sensors. 2008;**8**:3287

[127] Huang M, Wang Q, Zhang M, Zhu Q. Prediction of color and moisture content for vegetable soybean during drying using hyperspectral imaging technology. Journal of Food Engineering. 2014;**128**:24-30

[128] Maria T, Tsaniklidis G, Delis C, Nikolopoulou AE, Nikoloudakis N, Karapanos I, Aivalakis G. Gene transcript accumulation and enzyme activity of β-amylases suggest involvement in the starch depletion during the ripening of cherry tomatoes. Plant Gene. 2016;**5**:8-12

[129] Menesatti P, Zanella A, D'Andrea S, Costa C, Paglia G, Pallottino F, Zude M. Supervised multivariate analysis of hyper-spectral NIR images to evaluate the starch index of apples. Food & Bioprocess Technology. 2009;**2**:308-314

[130] Peirs A, Scheerlinck N, De Baerdemaeker J, et al. Starch index determination of apple fruit by means of a hyperspectral near infrared reflectance imaging system. Journal of Near Infrared Spectroscopy. 2003;**11**:379-389

[131] Chen WU, Jian-Guo HE, Xiao-Guang HE, et al. Non-destructive detection of starch content in potatoes based on near-infrared hyperspectral imaging technique. Journal of Henan University of Technology. 2014

[132] Trong NND, Tsuta M, Nicola BM, et al. Prediction of optimal cooking time for boiled potatoes by hyperspectral imaging. Journal of Food Engineering. 2011;**105**:617-624

[133] Crisosto CH. Stone fruit maturity indices: A descriptive review. Postharvest News & Information. 1994

[134] Garridonovell C, Pérezmarin D, Amigo JM, Fernándeznovales J, Guerrero JE, Garridovaro A. Grading and color evolution of apples using RGB and hyperspectral imaging vision cameras. Journal of Food Engineering. 2012;**113**:281-288

[135] Herrerolangreo A, Lunadei L, Lle L, et al. Multispectral vision for monitoring peach ripeness. Journal of Food Science. 2011;**76**:E178

[136] Girod D, Landry JA, Doyon G, Osuna-García JA, Salazar-García S, Goenaga R. Evaluating hass avocado maturity using hyperspectral imaging. Caribbean Food Crops Society. 2008

[137] Greensill C, Newman D. An investigation into the determination of the maturity of pawpaws (*Carica papaya*) from NIR transmission spectra. Journal of Near Infrared Spectroscopy. 1999;**7**:109-116

[138] Julio NB, José Miguel HH, Francisco José H. Determination of technological maturity of grapes and total phenolic compounds of grape skins in red and white cultivars during ripening by near infrared hyperspectral image: A preliminary approach. Food Chemistry. 2014;**152**:586-591

Hyperspectral Imaging for Assessing Quality and Safety of Meat

Wenxiu Wang and Yankun Peng

Additional information is available at the end of the chapter

http://dx.doi.org/10.5772/intechopen.74371

Abstract

Hyperspectral imaging (HSI) technology is a novel nondestructive method and has found various applications in the agricultural and food industry. In this chapter, the employment of HSI for meat quality assessment and safety control was summarized. The quality attributes include sensory attributes (color and marbling), chemical attributes (moisture, protein, intramuscular fat, and fatty acids), and technological attributes (pH, tenderness, and water holding capacity (WHC)). The safety attributes mainly include bacterial contamination and freshness determination. The spectral method is described in terms of the basic working principle, fundamental configurations, analysis period, and applications in meat assessment. In addition, the advantages, disadvantages, and problems to be tackled facing the HSI are also discussed. The current studies have demonstrated that HSI technology can be a potential tool to replace the traditional method for online and simultaneous evaluation of multiple quality and safety attributes of meat.

Keywords: hyperspectral imaging, meat, quality, safety, nondestructive detection

1. Introduction

Meat is an important element in human's diet as it is rich in easily digestible protein, vitamins, minerals, and micronutrients [1]. The improvement of living standards has prompted people to pay more attention to quality and safety, which are two crucial aspects related with meat. Quality refers to a comprehensive property about the sensory, chemical, and technological attributes of meat. The sensory attributes, such as color and marbling scores (MS), are about the first expression on meat and will influence consumers' willingness to purchase. Chemical attributes include moisture, protein, intramuscular fat (IMF), fatty acids (FAs), etc., and they are concerned with nutritive value and human health. Technological attributes like pH, water holding capacity (WHC), and tenderness are important factors that affect the eating quality [2]. Safety is another concern about meat, which can be challenged in several ways, such as biological issues and chemical contamination [3]. The former is considered as the biggest threat. This is because meat is ease of perishability and is suitable for the microbial growth.

The conventional methods for determination of the aforementioned traits include sensory evaluation and instrumental analysis [4]. They can provide accurate results and have been used for decades. However, they have some drawbacks as they are destructive, time-consuming, and labor-intensive. To match up with the demands of producers, manufacturers, distributors, retailers, and especially consumers for reliable and real-time evaluation of meat, nondestructive, rapid, and efficient tools are in urgent need. Hyperspectral imaging (HSI) technology has emerged as an alternative method. It combines the advantages of spectroscopic and imaging techniques and can acquire the spatial and spectral information simultaneously [5]. Owing to the merits, it has found numerous applications in agro-products for determination of internal traits and external features [6, 7]. A typical hyperspectral reflectance measurement system is shown in **Figure 1**, which mainly consists of a CCD camera, an imaging spectrograph, a zoom lens, illumination, computer, etc. When obtaining scattering images, an optical fiber is usually needed to form a point light source [8]. The hyperspectral data are collected in a three-dimensional (3D) form called "hypercube," among which two dimensions are spatial information expressed in x and y, and the third dimension is wavelength information, which is represented with λ [9].

After acquisition of reflectance images, chemometric tools are required to relate them and reference values of tested samples. Generally, the image is corrected and segmented first. Then, a range of interest (ROI) is selected and the mean spectrum is extracted from it. **Figure 2** shows the reflectance spectra extracted from a pixel. Subsequently, the spectrum undergoes (1) pretreatments to reduce and correct the adverse interferences caused by scattering, baseline drift, etc. The commonly used methods include Savitzky-Golay (SG)

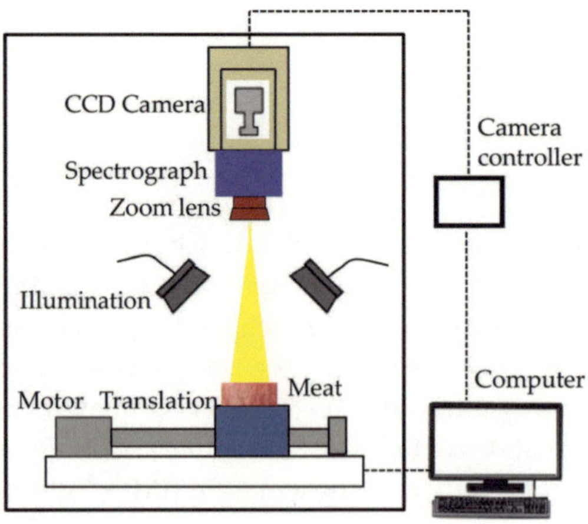

Figure 1. Schematic diagram of a hyperspectral reflectance measurement system.

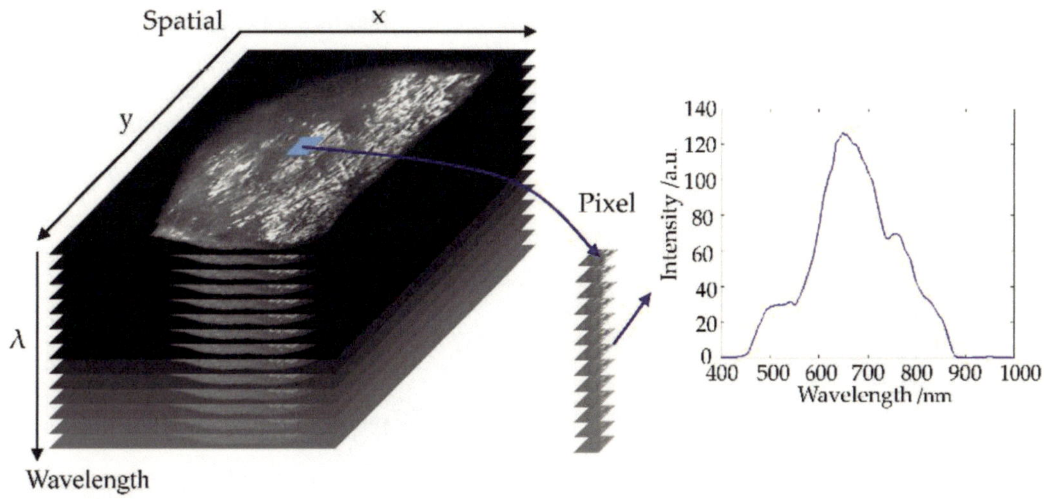

Figure 2. Reflectance spectra extracted from hyperspectral image.

smoothing, derivation, standard normal variate transformation (SNVT), orthogonal signal correction (OSC), and so on [10]; (2) feature wavelengths selection to eliminate the unwanted information and retain the characteristic wavelengths for simplifying the modeling. Genetic algorithm (GA), successive projection algorithm (SPA), uninformative variable elimination (UVE), etc. are efficient approaches for this purpose; (3) model establishment for quantitative or qualitative analysis using partial least squares regression (PLSR), multiple linear regression (MLR), least squares support vector machine (LS-SVM), and so on. In this case, chemical maps can be created by transferring the model to every pixel in the image to show distributions of each ingredient in a pixel-wise manner. Usually, the models are evaluated with correlation coefficient of calibration, prediction, and cross validation (R_c, R_p, and R_{cv}) and coefficient of determination for the calibration, prediction, and cross validation (R_c^2, R_p^2, and R_{cv}^2).

As to the spatially resolved hyperspectral images, effective nonlinear curve-fitting algorithms are usually required to extract scattering characteristics. Lorentzian distribution function is one of them, which has been intensively used in optic research to describe the light distribution patterns. It has three forms of expression: two-parameter, three-parameter, and four-parameter [11], and they are expressed as shown in Eqs. (1)–(3):

$$R = \frac{b}{1+(x/c)^2} \quad (1)$$

$$R = a+\frac{b}{1+(x/c)^2} \quad (2)$$

$$R = a+\frac{b}{1+(x/c)^d} \quad (3)$$

where R represents the light intensity, x is the scattering distance, a is the asymptotic value, b represents the peak value, c is the full width at $b/2$, and d is the slope around the inflection point. Similar with the Lorentzian distribution function, Gompertz function also has three forms of expression, namely, two-parameter, three-parameter, and four-parameter functions [11]. Eqs. (4)–(6) show their mathematical expressions:

$$R = 1 - \exp\{\exp(\varepsilon - \delta x)\} \tag{4}$$

$$R = 1 - (1 - \alpha)\exp\{-\exp(\varepsilon - \delta x)\} \tag{5}$$

$$R = \alpha + \beta[1 - \exp\{-\exp(\varepsilon - \delta x)\}] \tag{6}$$

where R represents the light intensity, x is the scattering distance, α is the asymptotic value, β is the upper value, ε represents the full scattering width, and δ is the slope value. Boltzmann function is another equation to describe light scattering and absorption in turbid materials. Absorption coefficient (μ_a) and scattering coefficient (μ_s') can be extracted to characterize the chemical or physical properties of tested samples. A more detailed introduction to the fitting functions was given by Peng [12].

As a vast number of data are contained in a hyperspectral image, the high-dimensional nature increases the difficulty in acquiring and processing the huge data. Multispectral imaging (MSI) is a simplified version, which uses few (generally less than 10) discrete spectral images. By acquiring the spatial and spectral information that are directly useful for meat detection, the experiment and analysis period are simplified [13]. Besides, simpler algorithms are needed, and the data size is decreased significantly, making it feasible to be implemented in the field or industry for real-time applications.

2. Quality assessment using HSI

2.1. Sensory quality attributes

Of all the sensory attributes, color is one of the most critical indicators which determine the first impression on meat for consumers. Meanwhile, they also reflect the freshness degree of meat and will in turn affect consumers' willingness to purchase. Color is related with the content and molecular type of myoglobin and hemoglobin in meat. Conventionally, the meat color is measured by means of a colorimeter to obtain the L^* (light), a^* (red-green), and b^* (yellow-blue). Marbling pattern is another important indicator, which would directly influence the grade and price of meat. The grading of meat is commonly conducted by well-trained professionals referring to different carcass grading standards.

To overcome the shortcomings of subjectivity and laboriousness in the traditional method, a HSI system in the spectral range of 400–1100 nm (**Figure 3a**) was used to acquire the scattering

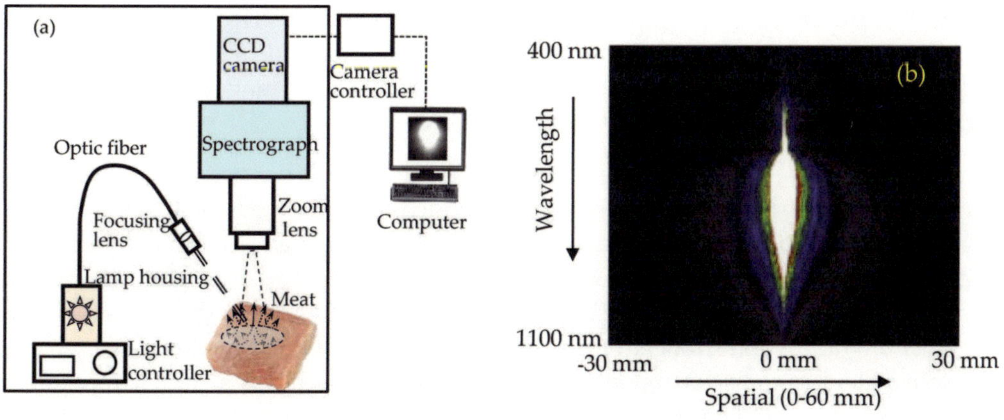

Figure 3. (a) Acquisition system for hyperspectral scattering image and (b) original scattering image for beef sample.

images of beef samples [14]. Different from that in **Figure 1**, a halogen tungsten lamp coupled with an optical fiber constituted the light source unit. The hyperspectral image was shown in **Figure 3b**, in which the vertical axis was wavelength axis and the horizontal axis represented the spatial distance. Then, three-parameter Lorentzian distribution function was employed to fit the scattering profile. A good fitting effect was observed between 450 and 1090 nm with the correlation coefficient greater than 0.90. The wavelengths below 450 nm and above 1090 nm contained considerable noise; hence, they were eliminated in the subsequent analysis.

Stepwise regression analysis was performed to determine the characteristic wavelengths before model establishment. Seven, seven, and eight wavelengths were selected for L^*, a^*, and b^*, respectively, as shown in **Table 1**. Most of the identified wavelengths were related with deoxymyoglobin, oxymyoglobin, metmyoglobin, and sulfmyoglobin. Then, MLR models based on the combined Lorentzian parameters were created with R_{cv} of 0.96, 0.96, and 0.97 for L^*, a^*, and b^*, respectively. The satisfactory results demonstrated the feasibility of spatially scattering information for color evaluation.

The nondestructive determination of meat color based on reflectance spectra was also carried out. Hyperspectral images of beef, lamb, and pork in the range of 400–1000 nm were collected [15]. Unlike the previous studies in which different feature wavelengths were selected for each attribute, a set of important variables were identified for L^*, a^*, and b^* to create models. They were 450, 460, 600, 620, 820, and 980 nm, which had little deviation from those chosen in

Quality attributes	Feature wavelengths (nm)
L^*	653, 678, 722, 868, 875, 920, 1050
a^*	465, 575, 614, 635, 671, 724, 978
b^*	486, 524, 540, 645, 700, 721, 780, 954

Table 1. Feature wavelengths for color selected by stepwise regression analysis.

Ref. [14]. The differences in image acquisition system and chemical compositions of meat may account for the deviation. Based on the six wavelengths, MLR models were built which gave R_p^2 of 0.97, 0.84, and 0.82 for L^*, a^*, and b^*. The results also laid foundation for the development of MSI detection device for color determination.

The capacity of HSI technology for beef marbling grade analysis has been confirmed. Images of a total of 33 beef samples were collected and divided into seven grades, 1, 1.5, 2, 2.5, 3, 3.5, and 4, according to the industry standard NY/T 676–2003 [16]. **Figure 4** showed the image of a sample with grade of 3 at 470, 550, 600, 660, 720, 850, and 950 nm. The eighth one in **Figure 4** was the combination of images at 720 (red), 550 (green), and 470 nm (blue). It can be seen that higher brightness and stronger muscle reflexes were found in the near-infrared region. In the visible region, the muscles and fat had larger contrast, which was favorable for marbling segmentation. Then, ROIs for the muscle and fat were selected, respectively, and their mean reflectance spectra were extracted from the ROI. Calculate the reflection intensity ratios of the muscle and fat at each wavelength, and the biggest ratio was observed at 530 nm. Hence, it was identified as the optimal wavelength for marbling feature extraction.

Based on the images at 530 nm, big fat area (higher than 14.88 mm^2 which corresponded to 75 pixels), medium fat area (between 3.72 and 14.88 mm^2), and small fat area (below 3.72 mm^2 which corresponded to 20 pixels) were calculated. Taking the three parameters as independent variables, MLR model was built with R_{cv}^2 of 0.92 and standard error of cross validation (SECV) of 0.45. Meanwhile, the regular decision function was also trialed to relate the image features and marbling grades with classification accuracy of 78.8%, which was lower than MLR model (84.8%). The overall results were satisfactory and demonstrated the feasibility of HSI for marbling evaluation.

The marbling scores of fresh, frozen, and frozen-thawed pork were quantified by a HSI system in the range of 900–1700 nm [17]. After ROI selection, a Gabor filter was performed

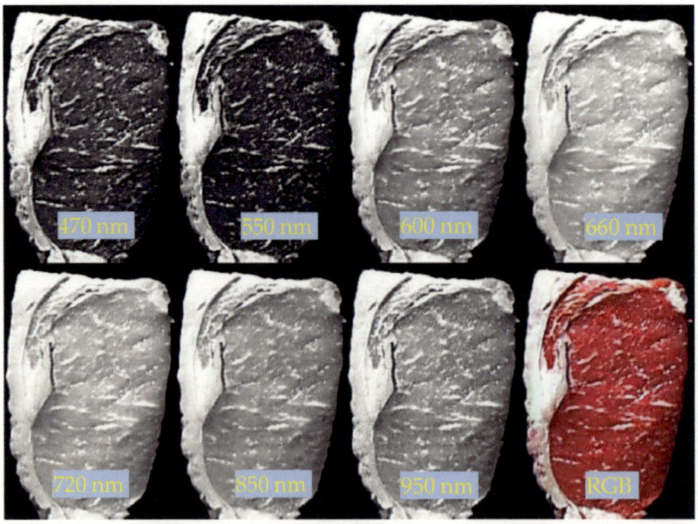

Figure 4. Images at seven wavelengths and RGB image of beef sample.

Sample	Image features	Key wavelengths	R_c	R_{cv}	R_p
Fresh	DMR	1082, 1188, 1217, 1236, 1452	0.85	0.83	0.91
Fresh	DMG	1082, 1188, 1236, 1346, 1380	0.88	0.86	0.88
Frozen	DMR	1217, 1236	0.84	0.83	0.90
Frozen	DMG	1217, 1231, 1264, 1514	0.83	0.83	0.85
Frozen-thawed	DMR	1169, 1255	0.82	0.81	0.89
Frozen-thawed	DMG	1078, 1174, 1226, 1346, 1433	0.89	0.87	0.91

Note: DMR and DMG represented the first derivative spectra of raw and Gabor-filtered images. R_c, R_{cv}, and R_p represented the correlation coefficient of calibration, cross validation, and prediction, respectively.

Table 2. Selected key wavelengths and model performance.

on the images, and the mean Gabor-filtered spectra and raw spectra were obtained. Then, they were pretreated with first derivative (FD), and MLR models were built linking the processed spectra at the feature wavelengths and marbling scores. The wavelengths selected and model results for fresh, frozen, and frozen-thawed pork were shown in **Table 2**. The promising results confirmed the capacity of Gabor filter technique in extracting image features.

2.2. Chemical quality attributes

Of all the chemical compositions, moisture, protein, and IMF are three major components of meat. FAs are also important indicators and have attracted more and more attentions in recent days. The reasons may be that the amount and types of FAs would influence other attributes and they also had some relationship with the cardiovascular diseases [18]. The traditional measurement methods for those attributes were drying method, Soxhlet method, Kjeldahl method, or gas chromatographic method. They are destructive, time-consuming, and laborious and cannot realize simultaneous detection of multiple attributes. As these chemical compositions contained C—H, N—H, or O—H bonds, the applications of HSI mainly focused on the usage of reflectance spectra to realize nondestructive determination.

A HSI system in the range of 900–1700 nm was used for determination of moisture, IMF, and protein in lamb meat [19]. A total of 126 samples containing three different muscles (semitendinosus (ST), semimembranosus (SM), and longissimus dorsi (LD)) were employed for hyperspectral image acquisition and reference analysis. After image correction and segmentation, spectral data were extracted from the ROIs. Samples were divided into calibration and validation set according to a ratio of 2:1, namely, 84 *vs.* 42 samples. The ranges of moisture, IMF, and protein in the calibration set were 69.65–76.35, 0.75–7.62, and 21.30–24.05%. In the prediction set, the ranges were 69.45–75.64, 0.74–6.01, and 21.24–23.84%. The moisture and IMF had a big standard deviation, while the protein had a relatively small standard deviation.

First, PLSR models based on full-band spectra were built. R_{cv}^2 of 0.94, 0.94, and 0.67 were found for moisture, IMF, and protein. The reason for the poor prediction ability for protein may be the narrow range of reference. To enhance the model for protein, some samples were minced with adjoined fat portion of the muscle to increase the reference ranges to 16.30–24.05%. New PLSR model was then built based on the modified data with R_{cv}^2 of 0.85, indicating a significant improvement. Feature wavelengths were further selected according to the regression coefficients obtained from the PLSR models. For moisture and IMF, 6 out of 237 wavelengths were selected, namely, 960, 1057, 1131, 1211, 1308, and 1394 nm. For protein, 1008, 1211, 1315, 1445, 1562, and 1649 nm were identified as the important wavelengths. Simplified models were again created with R_{cv}^2 of 0.86, 0.90, and 0.83 for moisture, IMF, and protein. The results were similar with those using the whole spectral range, which confirmed the capacity of HSI for prediction of chemical compositions.

As to the FA prediction using HSI, few studies were conducted. A HSI system in the near-infrared range of 1000–2300 nm was used to detect the FAs in intact raw beef cuts [20]. The FAs included the total saturated fatty acid (SFA), total unsaturated fatty acid (UFA), myristic (C14:0), palmitic (C16:0), stearic (C18:0), myristoleic (C14:1), palmitoleic (C16:1), oleic (C18:1), and linoleic (C18:2) acids. Similar with the aforementioned procedure, reflectance spectra were extracted and converted to absorbance values. Multiple scattering correction (MSC) was conducted on the absorbance spectra to correct spectral intensity differences. PLSR models were then built, yielding satisfactory results for SFA and UFA with R_p^2 of 0.87 and 0.89. For other attributes, the R_p^2 varied from 0.68 to 0.89. Apply the models to every pixel in the hyperspectral images, the chemical maps were generated, and the distributions of each composition can be observed intuitively.

2.3. Technological quality attributes

The technological quality attributes of meat mainly include pH, tenderness, and WHC, which are related to some structural and biochemical phenomena in living or carcass muscles. These attributes depend not only on the type of animal feed and fatty acid composition of carcass but also on the maturation effect. pH is considered as an important indicator of meat quality as it affects the color, tenderness, flavor after cooking, shelf life, and water retention. Meanwhile, it is also a reference to judge the meat freshness as it increases when meat turns spoiled. The traditional measurement for pH is by means of a pH meter to insert it into the meat. Tenderness directly affects the eating quality and commodity value of meat. It is closely related with the muscle structure (connective tissue) and biochemical composition (proteolysis of myofibrils and cytoskeletal proteins) of meat. Sensory evaluation and Warner-Bratzler shear force (WBSF) or slice shear force (SSF) are currently used methods for tenderness determination. WHC refers to the ability of the muscle to retain water and is an important factor that affects the color, flavor, and tenderness of meat. The conventional measurement methods include cooking loss method, drip loss method, squeezing method, and so on.

The determination of technological attributes using hyperspectral scattering imaging technique had been explored, and satisfactory results were obtained. Similar with the aforementioned analysis procedure for color, predictive models were built to relate the multiple "parameter spectra" of Lorentzian, Gompertz, or Boltzmann function and reference values. A sample of this topic about tenderness evaluation was given in Ref. [21]. A total of 31 pork

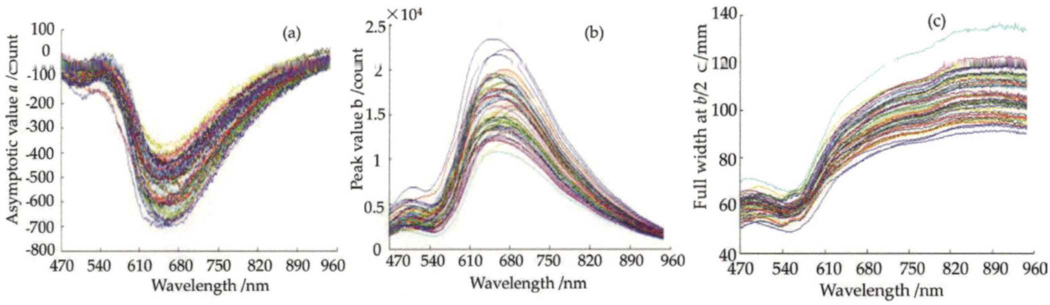

Figure 5. Lorentzian parameters extracted from pork images: (a) parameter *a*, (b) parameter *b*, and (c) parameter *c*.

samples were collected for scattering image acquisition and tenderness measurement by WBSF method. Then, the authors used three-parameter Lorentzian distribution function to fit the scattering profile, and accurate fitting performance was observed. For example, the fitting correlation coefficient at 575 nm was 0.998. Individual parameters *a*, *b*, and *c* (as shown in **Figure 5**) and the combination parameters of (*b*-*a*), (*b*-*a*) × *c*, (*b*-*a*)/*c*, and "*a*&*b*&*c*" were extracted. Likewise, stepwise regression analysis was performed to determine the optimal wavelength combinations for each parameter. Comparison of results showed that the models based on parameters *a*, *b*, (*b*-*a*), and (*b*-*a*)/*c* performed better with R_{cv} of 0.831, 0.860, 0.856, and 0.930, respectively.

Meanwhile, the modified Gompertz function was also employed to extract scattering characteristic of pork samples [22]. Promising fitting performance was found between 470 and 960 nm with coefficients all around 0.99. Parameters α, β, ε, and δ were then extracted, and their spectra at each wavelength were shown in **Figure 6**. As no optimal wavelengths were found for parameter β, hence, MLR models based on individual (α, ε, and δ) and integrated (α&ε&δ) were established and compared. The model based on the integrated one was superior to others with R_{cv} of 0.949, due to that more comprehensive information was involved. The overall results were better than the best result using Lorentzian parameter (*b*-*a*)/*c* (R_{cv} = 0.930).

Studies on using reflectance spectra in conjunction with multivariate analysis for noncontact measurement of pH, tenderness, and WHC have been conducted intensively. Hyperspectral images in the range of 900–1700 nm of beef samples were collected to predict WHC [23]. Samples were prepared with three different breeds and different muscles (*M. longissimus* dorsi (LD), M. semitendinosus (ST), and Psoas major (PM)). Thus, the reference values of WHC had a large variation, which was beneficial to build a robust model. PLSR model was then built to correlate the spectra and reference values measured by drip loss method, and an R_{cv}^2 of 0.89 was obtained. According to the regression coefficients of PLSR model, six important wavelengths at the peak positions were further selected, namely, 940, 997, 1144, 1214, 1342, and 1443 nm. New PLSR model based on the feature wavelengths was created with R_{cv}^2 of 0.87, demonstrating the potential of HSI for postmortem nondestructive determination of WHC. Key wavelengths for WHC in lamb meat in the range of 400–1000 nm were identified as 545, 610, 705, 765, 805, 900, 940, and 970 nm in Ref. [24]. Based on these wavelengths, LS-SVM model was built, yielding a good prediction performance with R_p^2 of 0.93.

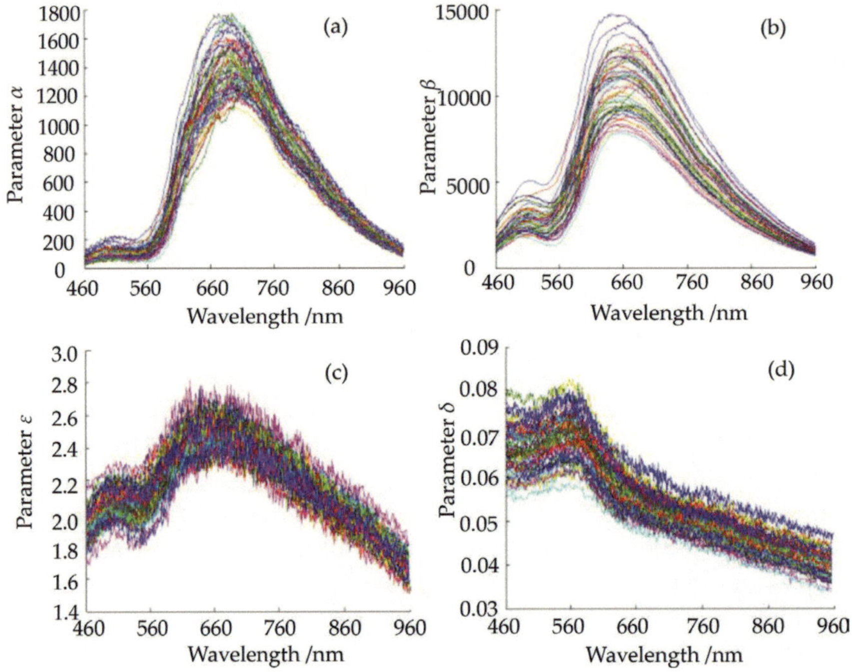

Figure 6. Gompertz parameters extracted from pork samples: (a) parameter α, (b) parameter β, (c) parameter ε, and (d) parameter δ.

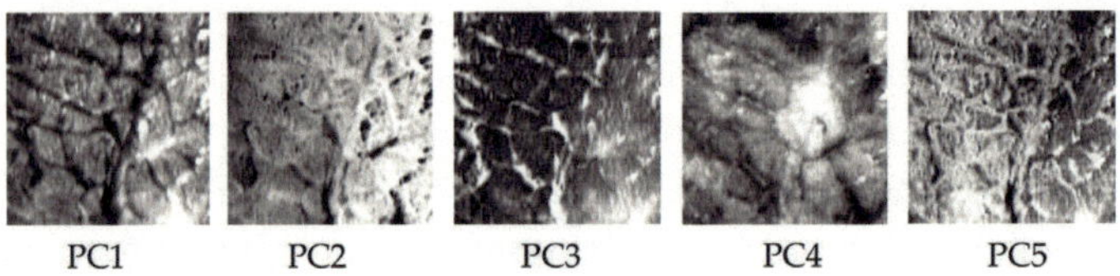

Figure 7. PC1–PC5 feature images for beef sample.

Besides, the feasibility of texture feature in predicting beef tenderness was also explored [25]. Three hundred and ninety-four hyperspectral images between 480 and 1020 nm were selected for each beef sample. Principal component analysis (PCA) was first conducted on these images to reduce the data dimension. The first five principal component images (PC1–PC5) are shown in **Figure 7**. It can be seen that the PC1 image contained most of the original image information with contribution rate more than 85%. The PC2 image contained little efficient information, while the PC3 image provided complementary information to PC1. The variance contribution rate of the first three principal component images was 95.37%, which was enough to represent the original information of the sample.

Further, gray-level co-occurrence matrix (GLCM) was used to extract texture variables from the three PCA images. Eight characteristic parameters including the mean, contrast, entropy,

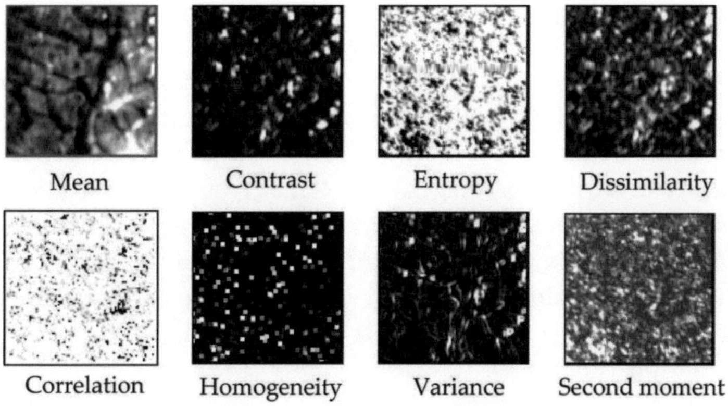

Figure 8. Eight texture features for image at 630 nm.

dissimilarity, correlation, homogeneity, variance, and second moment were obtained. **Figure 8** showed the texture features for the image at 630 nm. Thus, a total of 24 features were acquired for each sample. Taking them as the input variables, SVM and linear discrimination analysis (LDA) were established. The comparison of classification model results showed that LDA model had better performance than SVM with accuracy in the prediction set of 94.44%. The study also indicated that texture features can be used to determine the distribution of meat tenderness.

3. Safety control using HSI

Meat safety refers to the potential threat to human health. The applications of HSI on safety control mainly focused on the bacterial contamination and freshness determination. Bacterial contamination, often expressed as total viable counts (TVC), is an important microbiological indicator to determine the contamination and spoilage degree of meat. It is also used to predict the shelf life of meat or meat products. The quantity of 10^6 colony-forming units per gram (CFU/g) is considered as an acceptable limit, beyond which the meat became inedible [26]. Pseudomonas is a specific spoilage bacteria for meat stored at 4°C. The present method to evaluate TVC is plate-counting method, which is cumbersome and time-consuming and cannot satisfy the requirement for real-time detection. Freshness is a combination of flavor, taste, color, texture, and taste, and it is also a critical criterion to measure whether the meat meets the consumption standard. During the storage, proteins in meat tissue are broken down into low-molecular metabolites (alkaline substances such as ammonia and amines) with the action of enzymes and bacteria. Then, they are combined with acid within meat and form total volatile basic nitrogen (TVB-N) [27]. Traditional methods available for TVB-N detection include semimicro nitrogen determination and micro-diffusion method. They are time-consuming and destructive; hence, rapid and noncontact analytical methods are encouraged for TVB-N detection.

3.1. Bacterial contamination detection using HSI

In the previous studies, the ability of spatially scattering images for TVC, *Escherichia coli*, and Pseudomonas determination has been demonstrated. Such examples were given in Refs. [21, 28–29]. Hyperspectral images of beef samples stored within 2 weeks were collected [30]. To increase the signal-to-noise ratio of images, a 2 × 2 union (binning) operation was conducted, and then images of 520 × 688 pixels were acquired. For each sample, 4 different positions were selected for scanning and 4 times per position; thus, a total of 16 images were acquired for each sample. The reference values for TVC was then measured and recorded as $\log_{10}$CFU/g. **Figure 9** showed the bacterial growth curves. It can be seen that with storage time passed by, the TVC increased from 4.89 to 8.89.

Then, two-parameter Lorentzian distribution function was used to fit the scattering profiles. Parameters b, c, and $b \times c$ were obtained, and their correlation coefficients with $\log_{10}$CFU/g were shown in **Figure 10**. A similar trend was observed for the three parameters, especially for parameters b and $b \times c$, which had almost the same correlation coefficients over the range of 560–770 nm. Stepwise regression analysis was further performed to select the representative wavelengths for each parameter, as shown in **Table 3**. MLR models were established based on these wavelengths, and the results were also shown in **Table 3**. It can be seen that parameter $b \times c$ performed the best with R_p^2 of 0.96 and standard error of prediction set (SEP) of 0.23. The results demonstrated the feasibility of HSI for nondestructive determination of TVC.

The potential of reflectance spectra for bacterial contamination measurement has also been explored, and satisfactory results were obtained. Zheng et al. conducted a study to build a precise and simple model with low cost for TVC of pork [31]. Fifty chilled pork samples were collected and stored in a refrigerator at 4°C. Hyperspectral images in the range of 400–1100 nm were acquired, and spectra were extracted from a ROI and averaged. The original spectra for all the experimental samples were shown in **Figure 11**. Crests and troughs were observed between 530 and 580 nm, which were reported to be associated with myoglobin. The absorption peak at 980 nm was related to water absorption, corresponding to the second overtones of O—H stretching.

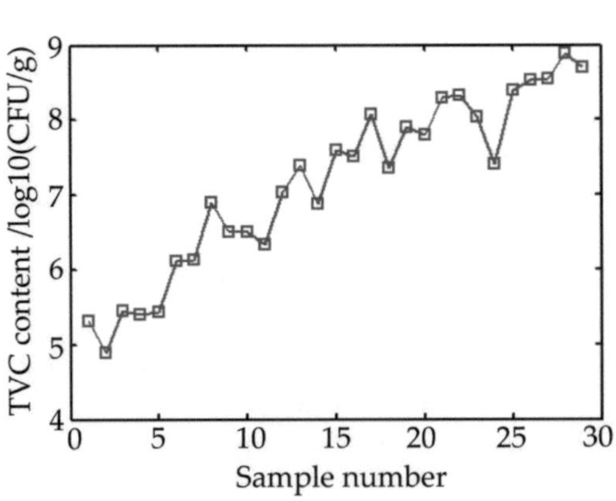

Figure 9. Bacterial growth curves during storage.

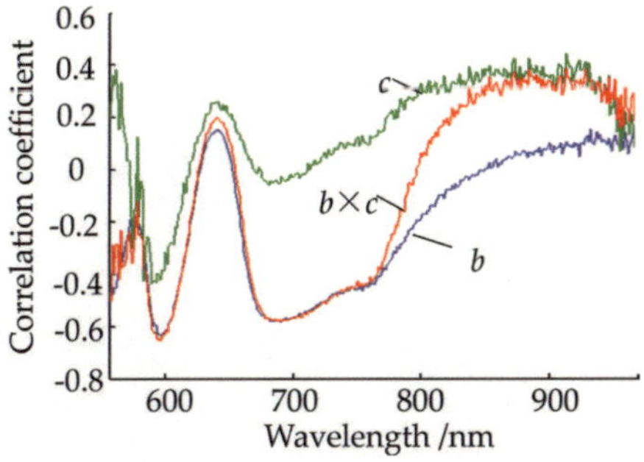

Figure 10. Correlation coefficients of parameters *b*, *c*, and $b \times c$ with $\log_{10}$CFU/g.

Parameter	Feature wavelengths (nm)	SEC	SEP	R_p^2	RSD (%)
b	592, 596, 602, 659, 803, 825	0.48	0.47	0.91	6.30
c	596, 838, 905, 913	0.70	0.62	0.69	8.31
$b \times c$	596, 822, 838, 841, 889, 900	0.44	0.23	0.96	3.08

Note: SEC, standard error of calibration set; SEP, standard error of prediction set; R_p^2, determination coefficient in the prediction set; RSD, relative standard deviation.

Table 3. Selected feature wavelengths and modeling results.

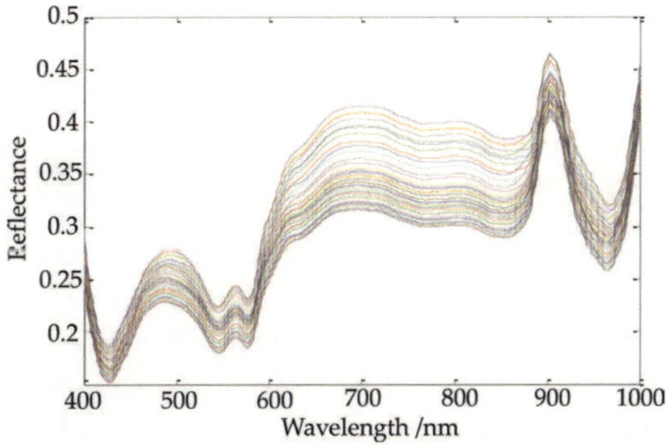

Figure 11. Original reflectance spectra of 50 pork samples.

To improve model accuracy, different pretreatment methods including SNVT, MSC, FD, second derivative (2D), Haar transformation (HT), and centering transformation (CT) were tried. Linear and nonlinear models, namely, PLSR and support vector regression (SVR) models, were built and compared. The SVR model combined with 2D pretreatment yielded the best result, with R_c and R_p of 0.99 and 0.94. Applying the optimal model to the hyperspectral images, the spatial distribution of bacteria can be observed clearly. **Figure 12** showed the chemical maps of TVC at different contamination levels, and the prediction results can be observed intuitively. It can be seen that there was an evident tendency of color change with

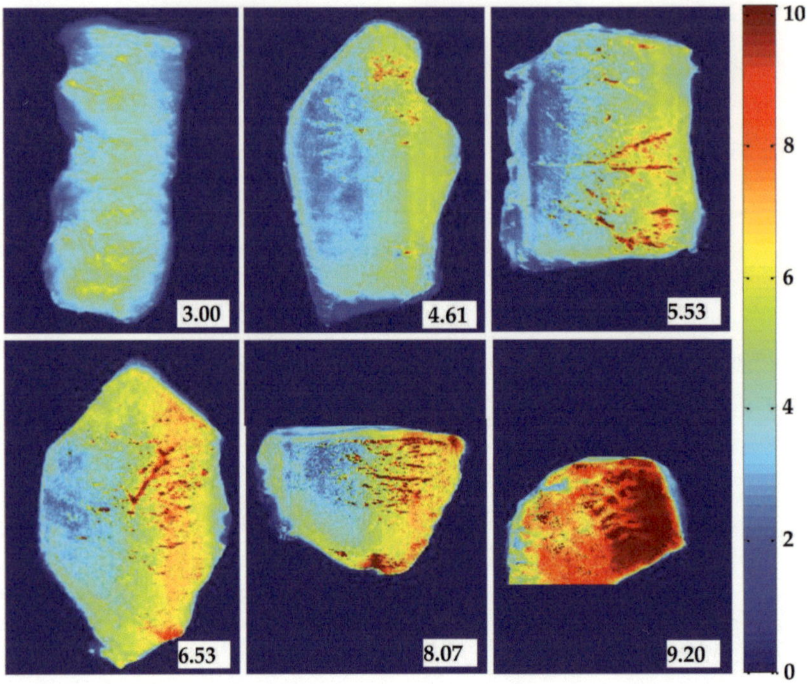

Figure 12. Visualizing maps of pork samples with different TVC values.

the increase of TVC content. When the bacterial contamination level was low, the density of blue color was dominant. In contrast, the red color occupied almost the whole image when the TVC content was 9.20. The change in color was inconsistent with the microbial content, which verified the prediction capacity of HSI for real-time monitoring of the bacterial contamination level.

In addition to these studies using HSI to predict TVC, MSI also demonstrated the application potential for TVC determination. One example was given in Ref. [32]. The authors used a rapid MSI device to detect pork spoilage with different storage temperatures (0, 5, 10, 15, and 20°C) and package types (aerobic and modified atmosphere). A reasonable prediction result for TVC was obtained with SEP of 7.47%, demonstrating the feasibility of using the setup to predict microbial counts in minced meat. Study for beef detection was carried out in Ref. [33]. The authors explored the capacity of MSI in determining microbial counts of aerobically packaged beef stored at different temperatures (0, 4, 8, 12, and 16°C). Average estimation deviations of 11.6, 13.6, and 16.7% were obtained for *Pseudomonas* spp., *B. thermosphacta*, and TVC, respectively. In recent days, "signature" spectra of contaminated aerobically packaged beef stored at 2, 8, and 15°C were extracted using MSI technique in Ref. [34]. According to a threshold of 2 $\log_{10}$CFU/g, samples were discriminated into two classes with accuracy of 80.8%.

3.2. Freshness determination using HSI

TVB-N is the most critical attribute related with freshness. According to the Chinese standard 2707–2016, meat is deemed to be semi-fresh or putrid if the TVB-N is beyond 15 mg/100 g.

As freshness is a complex criterion, generally, more than one attribute are combined to give a comprehensive evaluation of meat. The prediction ability of hyperspectral images within 470–1000 nm for simultaneous determination of TVB-N and pH was explored in Ref. [35]. Meat was classified into three grades: fresh (TVB-N ≤ 15 mg/100 g, pH 5.8–6.0), semi-fresh (TVB-N ≤ 25 mg/100 g, pH 6.0–6.3), and spoiled (TVB-N > 25 mg/100 g, pH > 6.3). Based on a halogen tungsten light source, hyperspectral images were acquired, as shown in **Figure 13a**. The vertical line represented the reflectance information at a certain position on the scanning line, and the horizontal line represented the diffusion information at a certain wavelength. Meat presented different diffusion characteristics at different wavelengths and positions. **Figure 13b** showed the spatial diffusion curves at 635, 760, 575, and 980 nm, and **Figure 13c** showed the reflectance spectra at 0, 5, 10, and 15 mm from the center of the scanning line. It can be seen that the spatial diffusion curve is a symmetrical pattern with a maximum intensity at the center of the scanning line. As the distance from the center increases at both ends, the intensity of the scattered light rapidly decreases. Based on the reflectance spectra pretreated with SG, TVB-N was well predicted with R_p of 0.90. According to the abovementioned classification standard, meat can be discriminated into different grades with total accuracy of 91%.

Another noteworthy report was given in Ref. [36]. The authors designed a portable device based on MSI technology to nondestructively detect TVB-N content in intact meat. The device was composed of hardware system, self-developed programming software, and the built-in prediction model. The schematic map of the developed portable device was shown in **Figure 14**, which mainly included an optical fiber, camera, filter, lens, computer, etc. The working flowchart was as follows: first, place the sample on the bracket, and press the external trigger switch, and then the image acquisition unit triggered the CCD camera to acquire image. Meanwhile, the filters were switched to obtain images at different wavelengths. After image collection, the software performed real-time processing, and the prediction results were displayed real time on the liquid crystal display. The scattering profiles of images were fitted using the four-parameter Lorentzian distribution function, and PLSR model based on the four parameters yielded a satisfactory result with R_p^2 of 0.87 for TVB-N.

Based on the images obtained by a MSI system, GLCM was also explored to extract feature variables for TVB-N prediction. In Ref. [37], features were obtained from three characteristic

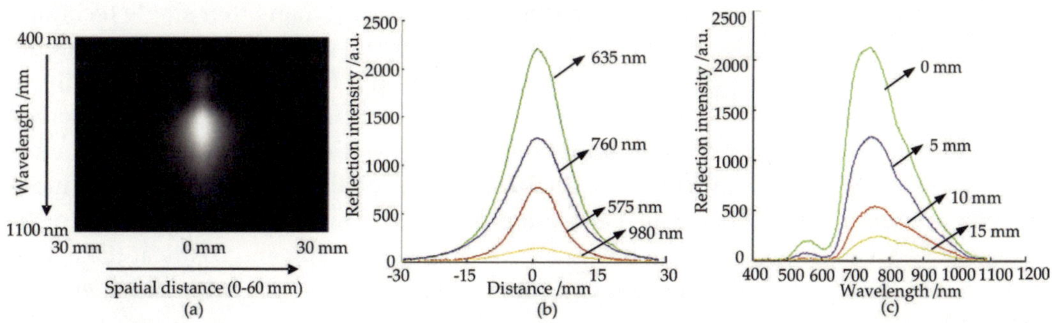

Figure 13. (a) Hyperspectral image of pork, (b) spatial diffusion curves at different wavelengths, and (c) reflectance spectra at different positions.

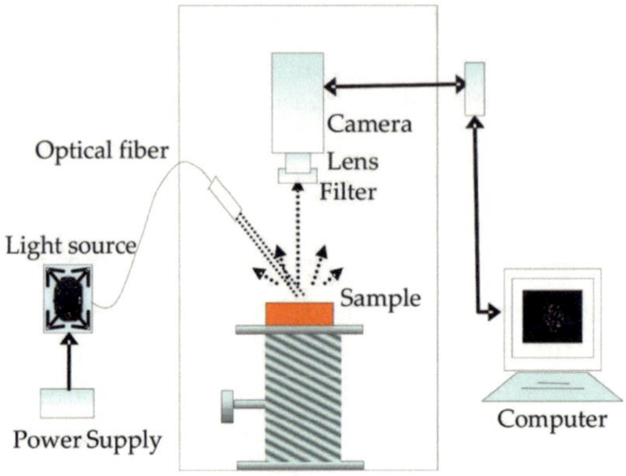

Figure 14. Portable device for TVB-N determination based on MSI.

images at 1280, 1440, and 1660 nm. Then, a new algorithm, namely, back propagation neural network adaptive boosting method, was proposed, obtaining a reasonable prediction result R_p of 0.8325. Their work indicated that the MSI system can be an efficient tool for TVB-N determination, and the research would facilitate its practical usage in meat industry.

4. Conclusions

This chapter summarized the applications of HSI and MSI in quality evaluation and safety control of meat. The current studies have demonstrated the capacity of HSI in quantitatively and qualitatively detection of meat. By integrating both spectroscopic and imaging techniques in one system, the spectral and spatial information of tested samples are acquired simultaneously, which paves way for its extensive applications in meat assessment. As more detailed and comprehensive characteristics of meat are extracted, the quality and safety attributes can be better predicted than using single spectroscopic or imaging technology.

However, despite the fact that the HSI and MSI technology has gained significant development, there are still several drawbacks facing this technology. First, as spectral and spatial information are collected simultaneously, a huge data are acquired for one scanning. Thus, image acquisition and processing become more difficult than other single technologies. With such massive raw image data, it is difficult for HSI systems to be widely implemented for online and real-time application. Secondly, the HSI instrument is relatively expensive compared with conventional methods, thus increasing the cost of commercial detection and impeding its broader adoption. Thirdly, as the nature of nondestructive prediction for HSI technology lies in the establishment of qualitative or quantitative models with reference values, laborious calibration procedures are necessary. Much effort is needed to build accurate models, which makes it time-consuming and costly at the beginning. In addition, the subsequent model updating also requires substantial time, energy, and funds.

To overcome these difficulties, some effort has been taken as possible solutions in response to these disadvantages. On the one hand, to speed up the analysis period, identifying the most influential wavelengths to eliminate the irrelevant information is an efficient method. In fact, various algorithms for feature wavelengths selection have been proposed for elimination of redundant information and reduction of multicollinearity problem. However, the important variables chosen by different approaches are not consistent even for the same set of spectra, and some selected wavelengths lack of scientific interpretability. Hence, more efficient chemometric methods are in need to improve model performance and robustness. On the other hand, the hardware system with good performance is the precondition and foundation of obtaining stable and high signal-to-noise optical signals. Hence, the enhancement in instrumental development in combination with the availability of high-speed computer will facilitate this technique to be dominant in the future. With further research and development, the HSI and MSI technology can become a powerful tool for online and real-time monitoring the quality and safety of meat.

Acknowledgements

The authors gratefully acknowledge the Special Fund for Agro-scientific Research in the Public Interest Program (Project No. 201003008) and the National Key Research and Development Program (Project No. 2016YFD0401205).

Conflict of interest

We declare that we have no conflict of interest.

Author details

Wenxiu Wang and Yankun Peng*

*Address all correspondence to: ypeng@cau.edu.cn

China Agricultural University, Beijing, China

References

[1] Wojnowski W, Majchrzak T, Dymerski T, Gebicki J, Namiesnik J. Electronic noses: Powerful tools in meat quality assessment. Meat Science. 2017;**131**:119-131. DOI: 10.1016/j.meatsci.2017.04.240

[2] Miller RK. The eating quality of meat. In: Fidel T, editor. Lawrie's Meat Science. 8th ed. Holland: Elsevier; 2017. pp. 461-499. DOI: 10.1016/b978-0-08-100694-8.00015-7

[3] Saucier L. Microbial spoilage, quality and safety within the context of meat sustainability. Meat Science. 2016;**120**:78-84. DOI: 10.1016/j.meatsci.2016.04.027

[4] Peng YK, Dhakal S. Optical methods and techniques for meat quality inspection. Transactions of the Asabe. 2015;**58**:1371-1386. DOI: 10.13031/trans.58.11004

[5] Kamruzzaman M, Barbin D, ElMasry G, Sun DW, Allen P. Potential of hyperspectral imaging and pattern recognition for categorization and authentication of red meat. Innovative Food Science & Emerging Technologies. 2012;**16**:316-325. DOI: 10.1016/j.ifset.2012.07.007

[6] Xiong ZJ, Sun DW, Zeng XA, Xie AG. Recent developments of hyperspectral imaging systems and their applications in detecting quality attributes of red meats: A review. Journal of Food Engineering. 2014;**132**:1-13. DOI: 10.1016/j.jfoodeng.2014.02.004

[7] He HJ, Sun DW. Hyperspectral imaging technology for rapid detection of various microbial contaminants in agricultural and food products. Trends in Food Science & Technology. 2015;**46**:99-109. DOI: 10.1016/j.tifs.2015.08.001

[8] Peng YK, Lu RF. Analysis of spatially resolved hyperspectral scattering images for assessing apple fruit firmness and soluble solids content. Postharvest Biology and Technology. 2008;**48**:52-62. DOI: 10.1016/j.postharvbio.2007.09.019

[9] Kamruzzaman M, Makino Y, Oshita S. Non-invasive analytical technology for the detection of contamination, adulteration, and authenticity of meat, poultry, and fish: A review. Analytica Chimica Acta. 2015;**853**:19-29. DOI: 10.1016/j.aca.2014.08.043

[10] Porep JU, Kammerer DR, Carle R. On-line application of near infrared (NIR) spectroscopy in food production. Trends in Food Science & Technology. 2015;**46**:211-230. DOI: 10.1016/j.tifs.2015.10.002

[11] Peng YK, Lu RF. Prediction of apple fruit firmness and soluble solids content using characteristics of multispectral scattering images. Journal of Food Engineering. 2007;**82**(2):142-152. DOI: 10.1016/j.jfoodeng.2006.12.027

[12] Peng YK. Spectral scattering for assessing quality and safety of meat. In: Lu RF, editor. Light Scattering Technology for Food Property, Quality and Safety Assessment. 1st ed. Boca Raton: CRC Press; 2016. pp. 283-317. DOI: 10.1201/b20220-12

[13] Liu J X, Cao Y, Wang Q, Pan W J, Ma F, et al. Rapid and non-destructive identification of water-injected beef samples using multispectral imaging analysis. Food Chemistry. 2016;**190**:938-943. DOI: http://dx.doi.org/10.1016/j.foodchem.2015.06.056

[14] Wu JH, Peng YK, Li YY, Wang W, Chen JJ, et al. Prediction of beef quality attributes using VIS/NIR hyperspectral scattering imaging technique. Journal of Food Engineering. 2012;**109**:267-273. DOI: 10.1016/j.jfoodeng.2011.10.004

[15] Kamruzzaman M, Makino Y, Oshita S. Online monitoring of red meat color using hyperspectral imaging. Meat Science. 2016;**116**:110-117. DOI: 10.1016/j.meatsci.2016.02.004

[16] Gao XD, Wu JH, Peng YK, Chen JJ, Tao FF. Analysis of beef-marbling grade using hyperspectral imaging technology. Academic Periodical of Farm Products Processing. 2009;**10**:33-37. DOI: 10.3969/j.issn.1671-9646(X).2009.10.008

[17] Huang H, Liu L, Ngadi MO. Prediction of pork fat attributes using NIR images of frozen and thawed pork. Meat Science. 2016;**119**:51-61. DOI: 10.1016/j.meatsci.2016.02.042

[18] Gonzalez-Martin I, Gonzalez-Perez C, Alvarez-Garcia N, Gonzalez-Cabrera JM. On-line determination of fatty acid composition in intramuscular fat of Iberian pork loin by NIRs with a remote reflectance fibre optic probe. Meat Science. 2005;**69**:243-248. DOI: 10.1016/j.meatsci.2004.07.003

[19] Kamruzzaman M, ElMasry G, Sun DW, Allen P. Non-destructive prediction and visualization of chemical composition in lamb meat using NIR hyperspectral imaging and multivariate regression. Innovative Food Science & Emerging Technologies. 2012;**16**:218-226. DOI: 10.1016/j.ifset.2012.06.003

[20] Kobayashi K-I, Matsui Y, Maebuchi Y, Toyota T, Nakauchi S. Near infrared spectroscopy and hyperspectral imaging for prediction and visualisation of fat and fatty acid content in intact raw beef cuts. Journal of Near Infrared Spectroscopy. 2010;**18**(5):301-315. DOI: 10.1255/jnirs.896

[21] Tao FF, Peng YK, Li YY, Chao KL, Dhakal S. Simultaneous determination of tenderness and Escherichia Coli contamination of pork using hyperspectral scattering technique. Meat Science. 2012;**90**:851-857. DOI: 10.1016/j.meatsci.2011.11.028

[22] Tao FF, Peng YK. A method for nondestructive prediction of pork meat quality and safety attributes by hyperspectral imaging technique. Journal of Food Engineering. 2014;**126**:98-106. DOI: 10.1016/j.jfoodeng.2013.11.006

[23] ElMasry G, Sun DW, Allen P. Non-destructive determination of water-holding capacity in fresh beef by using NIR hyperspectral imaging. Food Research International. 2011;**44**:2624-2633. DOI: 10.1016/j.foodres.2011.05.001

[24] Kamruzzaman M, Makino Y, Oshita S. Hyperspectral imaging for real-time monitoring of water holding capacity in red meat. LWT-Food Science and Technology. 2016;**66**:685-691. DOI: 10.1016/j.lwt.2015.11.021

[25] Zhao J, Peng YK. Distribution of beef tenderness grading based on texture feature by hyperspectral image analysis. Transaction of the Chinese Society for Agricultural Machinery. 2015;**31**:279-286. DOI: 10.3969/j.issn.1002-6819.2015.07.039

[26] Tao FF, Peng YK. A nondestructive method for prediction of total viable count in pork meat by hyperspectral scattering imaging. Food and Bioprocess Technology. 2014;**8**(1):17-30. DOI: 10.1007/s11947-014-1374-y

[27] Li HH, Chen QS, Zhao JW, Wu MZ. Nondestructive detection of total volatile basic nitrogen (TVB-N) content in pork meat by integrating hyperspectral imaging and colorimetric sensor combined with a nonlinear data fusion. LWT-Food Science and Technology. 2015;**63**:268-274. DOI: 10.1016/j.lwt.2015.03.052

[28] Tao FF, Wang W, Li YY, Peng YK, Wu JH. A rapid nondestructive measurement method for assessing the total plate count on chilled pork surface. Spectroscopy and Spectral Analysis. 2010;**30**:3405-3409. DOI: 10.3964/ j .issn.1000-0593(2010)12-3405-05

[29] Zhang LL, Peng YK. Noninvasive qualitative and quantitative assessment of spoilage attributes of chilled pork using hyperspectral scattering technique. Applied Spectroscopy. 2016;**70**:1309-1320. DOI: 10.1177/0003702816654060

[30] Peng YK, Zhang J, Wang W, Li YY, Wu JH, Huang H, et al. Potential prediction of the microbial spoilage of beef using spatially resolved hyperspectral scattering profiles. Journal of Food Engineering. 2011;**102**(2):163-169. DOI: 10.1016/j.jfoodeng.2010.08.014

[31] Zheng XC, Peng YK, Wang WX. A nondestructive real-time detection method of total viable count in pork by hyperspectral imaging technique. Applied Sciences. 2017;**7**(3):213. DOI: 10.3390/app7030213

[32] Dissing BS, Papadopoulou OS, Tassou C, Ersbøll BK, Carstensen JM, et al. Using multispectral imaging for spoilage detection of pork meat. Food and Bioprocess Technology. 2012;**6**:2268-2279. DOI: 10.1007/s11947-012-0886-6

[33] Panagou EZ, Papadopoulou O, Carstensen JM, Nychas GJ. Potential of multispectral imaging technology for rapid and non-destructive determination of the microbiological quality of beef filets during aerobic storage. International Journal of Food Microbiology. 2014;**174**:1-11. DOI: 10.1016/j.ijfoodmicro.2013.12.026

[34] Tsakanikas P, Pavlidis D, Panagou E, Nychas GJ. Exploiting multispectral imaging for non-invasive contamination assessment and mapping of meat samples. Talanta. 2016;**161**:606-614. DOI: 10.1016/j.talanta.2016.09.019

[35] Zhang LL, Li YY, Peng YK, Wang W, Jiang FC, et al. Determination of pork freshness attributes by hyperspectral imaging technique. Transactions of the Chinese Society of Agricultural Engineering. 2012;**28**:254-259. DOI: 10.3969/j.issn.1002-6819.2012.07.042

[36] Li CL, Peng YK, Tang XY. Device for rapid nondestructive detection of pork freshness based on multispectral imaging technology. Transactions of the Chinese Society of Agricultural Machinery. 2012;**43**:202-206. DOI: 10.6041/j.issn.1000-1298.2012.S0.040

[37] Huang QP, Chen QS, Li HH, Huang GP, Ouyang Q, Zhao JW. Non-destructively sensing pork's freshness indicator using near infrared multispectral imaging technique. Journal of Food Engineering. 2015;**154**:69-75. DOI: 10.1016/j.jfoodeng.2015.01.006

Chapter 5

Models Fitting to Pattern Recognition in Hyperspectral Images

Uziel Francisco Grajeda-González,
Alejandro Isabel Luna-Maldonado,
Humberto Rodriguez-Fuentes,
Juan Antonio Vidales-Contreras,
Ernesto Alonso Contreras-Salazar and
Héctor Flores-Breceda

Additional information is available at the end of the chapter

http://dx.doi.org/10.5772/intechopen.73159

Abstract

Worldwide, the concern on food safety, for example, on agriculture products, has become a topic with huge relevance. Nowadays, hyperspectral imaging systems for rapid detection of dangerous agents have emerged in response to these needs. In this research project, we proposed a new algorithm for *Salmonella typhimurium* detection on tomato surfaces in visible range (400–1000 nm). Gaussian model was used as a way to take out a model that could be calculated its definite integral; the final result of this algorithm is the area under curve (AUC), which gives a quantitative approach of spectral signatures. Three doses (5, 10, and 15 μL) and a control response (0 μL) were spread out on 20 tomatoes' surface. Subsequently, it was observed that some decrease responses with higher dose; also, numerically this pattern was seen with the help of AUC value. As a last step, a single factor analysis of variance showed no significance due to doses. Despite this outcome, the algorithm provides to be a good methodology for pathogen detection.

Keywords: food safety, hyperspectral imaging system, AUC, Gaussian model, ANOVA

1. Introduction

Hyperspectral imaging technology has been well developed in different areas of the industry, such as mining, quality assessment in food processes and detection of diseases that affect crops and fruits, among others [1]; it is also important to mention that nowadays the reduction cost of sensors and electrical circuits has allowed the gradual immersion of hyperspectral imaging systems. Supervised learning is characterized by the need to know the expected responses based on human knowledge or the characteristics of the system; these responses are known as target function; then the system tries to compare our inputs (in our case the set of pixels) with this function, to the process of comparison, and testing of inputs with expected responses is called learning; the learning process ends when the algorithm has an acceptable level of performance; supervised learning can be grouped into two approaches such as classification and regression [2].

Classification approach: the data should be grouped into "categories," for example, "infected," "uninfected," "damaged," "undamaged," and "mature."

Regression approach: data are treated as continuous function that can be modeled with mathematical functions that predict behavior. Some examples of supervised machine learning algorithms are:

- Linear regression
- Support vector machines
- Supervised neuronal network

On the other hand, the unsupervised processes try to model the distribution of the data and thus to obtain conclusions; this type of algorithm group has similar characteristics along the data by itself, without the help of expected knowledge [3]. Accordingly, there are two approaches:

Clustering: in this type of analysis, the result is groups of data that share characteristics associated with certain trends, for example, the economy of a country with respect to the level of education of its population.

Association: in this type of analysis you want to find rules that describe a large portion of the data, such as "people who buy X also tend to buy Y."

Examples of these unsupervised learning algorithms are:

- PCA
- PLSR
- Fisher's discriminant analysis
- K-means clustering
- Unsupervised neuronal network

Choosing one of these methodologies to work depends on conditions of the experiment, that is, if into the experiment, the possibility to calibrate the algorithm with an expected response exists, for example, if an expert is able to detect damaged areas in a crop, before starting the research, this information could generate the target function and can be used to train the

http://dx.doi.org/10.5772/intechopen.73159

algorithm [4]. Besides, data provided for hypercubes are usually analyzed by statistical pattern recognition approaches in three-dimensional space; these analyses come across from the simplest to the most complex; an additional way to getting relevant information from spectral data is analyzing its shape with curve fitting. Also, hyperspectral curve fitting methodology has the advantage of modeling multiple overlapping absorption, transmittance, and reflectance, with substantial less bands [5].

Moreover, wavelet is another technique that has impacted the way to analyze hyperspectral data. Due to its application on fields such as signal and image processing, pattern recognition, and data compression, wavelet transform has been an alternative for data analysis and dimensionality reduction [6]; the main idea of processing with wavelet transform is to decompose a signal into a series of shifted and scaled sub-representation of the mother wavelet function. This decomposition provides a hierarchical framework for interpreting the spectral information; some researchers have utilized wavelet transform for feature extraction, for example, classification of health and damage areas in leaves [7]. Other researchers [8] have studied the combination between PCA and wavelet coefficient to improve dimension reduction, and also they could highlight the small variations contained in spatial information. Another interesting application performed with wavelets was the fusion between hyperspectral and multispectral data [9]; the fusion image that proved to have more relevant information due to wavelets could be considered as a low-pass and high-pass filters that allow separate information which is not found with the naked eye.

Several researches that worked with modeling fitting and wavelet approach can be found in scientific literature; we are going to mention some of them: in Ref. [10], they investigated anomaly detection on a test data cube taken from a part of San Diego International Airport; in this research they proposed to use a Gauss-Markov algorithm to detect and classify statistical parameters within the data, that is, covariance matrix; as a result they show two binary images with 100% of target detection. It was developed a new algorithm [11] based on index total chlorophyll (C_{ab}) content; they proposed a new index called area under curve normalized to maximal band depth between 650 and 725 nm ($ANMB_{650–725}$); as a preliminary step, the area under the continuum-removed reflectance curve in the range of 650–725 nm ($AUC_{650–725}$) was computed. As an outcome, using area under curve (AUC) divided by a maximal band depth could predict chlorophyll content with good accuracy. It should be noted that despite the fact most of the current equipment operates between 400 and 2500 nm (visible and near infrared), it is important to select correctly the bands which contain the data where the area of interest is located; due to this fact, numerous works that focus on algorithms for band selection exist [12–14]. In addition, Ref. [15] compared different mathematical models for describing the hyperspectral scattering data in order to predict fruit firmness and soluble solid content (SSC) of Golden Delicious apples; the model utilized in the research was the Lorentzian distribution function, which gave a high fitting with an average correlation coefficient (r) greater than 0.995, owing to the oval shape of apples; it was necessary to calculate the integral of the measurement reflectance as a function of the area covered by the lens of the camera and the reflectance intensity I over the acceptance angle. As a conclusion, they mention that mathematical modeling of scattering data to obtain the total light reflectance, using an appropriate Lorentzian function, can provide a good way to predict apple fruit firmness and soluble solid content.

2. *Salmonella typhimurium* detection using hyperspectral imaging system

Foodborne detection has been a topic of interest in recent decades, due to food industry and government regulations. Traditional techniques based on agar culture media have huge shortcomings in rapid confirmation response and the inability to analyze a large number of samples; another disadvantage is the need to destroy the fruits in order to carry out the planting on the culture media. Moreover, hyperspectral imaging system has emerged as tool to detect bacteria in a considerable reduced time [16].

Specifically, *S. typhimurium* infection is usually transmitted by consumption of contaminated fruits, vegetables, fresh beef, or pork. Outbreaks caused by these bacteria have been reported in Canada, Europe, and the United States [17]; the symptoms of these bacteria are gastrointestinal problems, fevers, and in some cases death.

On the other hand, Mexican tomato production faces the challenge of complying with regulations imposed by the United States (USA) and Canada, where agricultural products must comply with safety features for sale in the foreign market; as well as economic losses in recent years due to the waiting for a long period of time and doubtful detection on infectious agents have caused the need for faster and more efficient detection methods [18].

This research project was focused on on the obtaining of the hyperspectral signatures and Gaussian prediction models with high fitting to calculate the AUC and with this information detect *S. typhimurium* on tomato surface. Hyperspectral imaging system promises to be a good technique for worldwide food safety.

3. Materials and methods

3.1. Biotechnological material

S. typhimurium was used, suspended in a media culture (broth in cryopreservation state) necessary for its survival. The experiment utilized commercial selective media *Salmonella-Shigella* (SS) agar, Hektoen enteric, and xylose lysine deoxycholate (XLD) agar. To isolate the bacteria, the streak plate isolation method was used; to display and select the suspect colonies more easily, this procedure was performed in triplicate. Assay tubes with 5 mL of tetrathionate broth were inoculated with *S. typhimurium* strain. This culture medium contains peptone and sodium carbonate, and the selectivity is the result of the presence of sodium thiosulfate that generates tetrathionate when added at a ratio of 0.2% iodine-iodide solution and 0.1% of bright green to each tube, allowing the growth of bacteria containing the reductase enzyme tetrathionate, and inhibits the development of other accompanying microorganisms. The incubation time was 24 h at a temperature of 37°C under aerobic conditions.

3.2. Tomato samples

Twenty tomatoes (*Solanum lycopersicum* L.) variety "Roma" were selected in a state of postharvest; they were purchased at a local supermarket in the municipality of General Escobedo, Nuevo Leon, Mexico. The tomatoes complied with high visual quality.

http://dx.doi.org/10.5772/intechopen.73159

3.3. Hypercube of contaminated tomato

3.3.1. Hyperspectral system

The hyperspectral equipment utilized for this research was the PIKE F210b (Alliend Vision Technologies, GmbH); the camera is coupled to a Spectograph ImSpector V10E (Specim, Spectral Imaging Ltd.); the hyperspectral system is attached to a linear translation structure, which is, essentially, a band, a motor, and a speed regulation stage. This is necessary due to the push-broom operation; besides, the spectral range of the equipment goes from 400 to 1000 nm. Finally, the system works with two halogen-tungsten bulbs with a power of 60 W.

3.3.2. Sample inoculation

In order to start the research, the first step was inoculating the surface of 20 tomatoes with *Salmonella typhimurium* bacteria at three different amounts of dosification; these were 5, 10, and 15 μL and a zone with no contamination (0 μL), as we can see in **Figure 1**. The spread of a little drop on the tomato surface it was carry out with the help a micropipette.

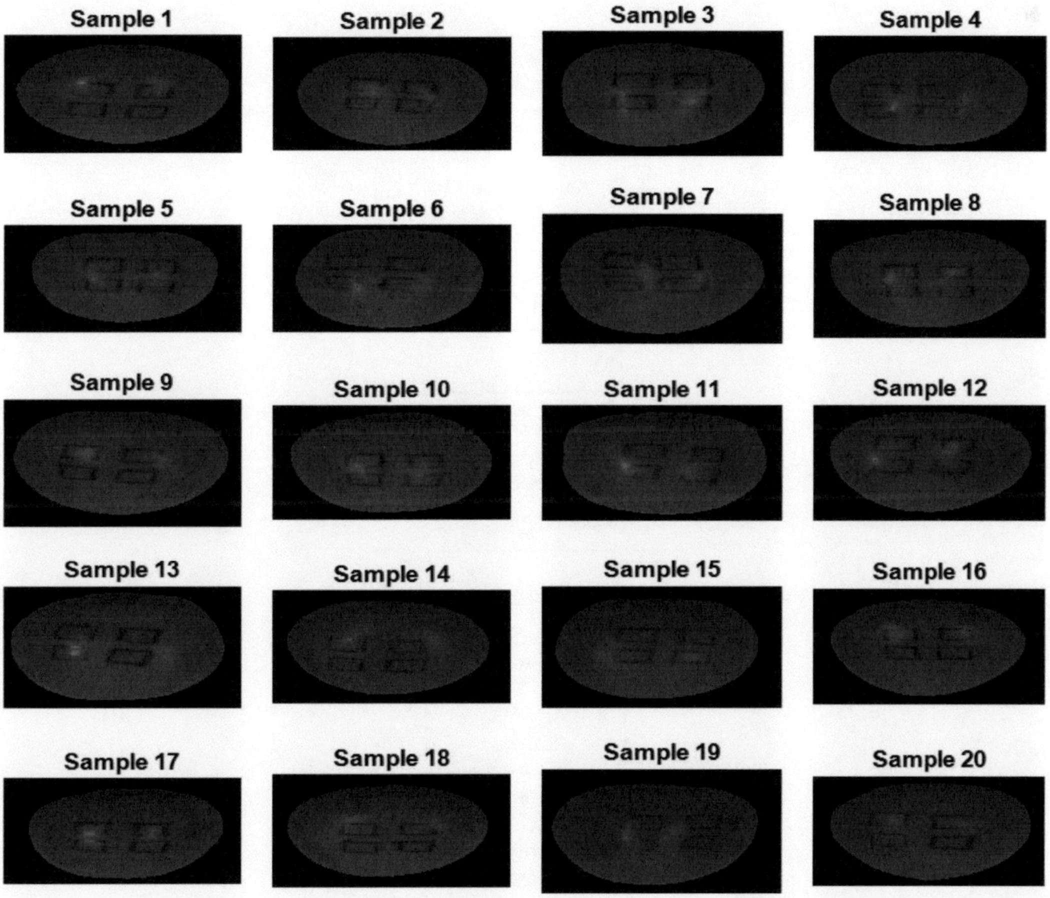

Figure 1. Tomato hypercubes shown in wavelength 692 nm; each tomato was labeled with four square-shaped zones which was spread with 0, 5, 10, and 15 μL of *S. typhimurium*.

20 hypercubes were obtained with a 600 × 1920 spatial resolution and 1080 bands with 12 bits of resolution.

3.3.3. Preprocessing and data preparation

Hyperspectral imaging processing usually has a pre-stage called preprocessing, necessary to remove the effect of death pixels, noisy signals, errors caused by analog to digital process conversion, etc. Additionally, due to high abundance of data, it required a calibration process and test hypercubes for correcting data [19]. A general workflow is shown in **Figure 2**, and its subsequent analysis is discussed in the next section.

3.3.4. Normalization

The analysis of hypercubes involves huge amount of data, thence one of the main reasons for be adapt the hypercubes to more manageable sizes and with this to improve the computer processing time.

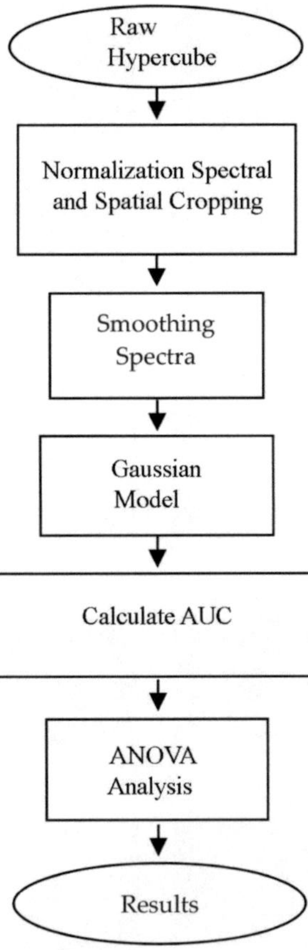

Figure 2. Workflow of proposed algorithm for pathogen detection, using area under curve.

As the first step, normalization of all hypercube was carried out using Eq. (1):

$$Hypercube_{Normalized} = \frac{RawHypercube - Dark\_reference}{White\_reference - Dark\_reference} \quad (1)$$

where $Hypercube_{Normalized}$ is the calibrated hypercube, *RawHypercube* is the total data without any type of process, *Dark_reference* was taken with the absence of illumination and camera lens covered, and *White_reference* data cube was generated with a high reflectance white mosaic and the lights on.

3.3.5. Spatial and spectral crop

As was mentioned before, each cube of data had a spatial dimension of 600 × 1920. It should be noted that most of this information is merged with the background, which is not necessary to analyze, from there that a spatial cropping was necessary. Each cube was reduced to an average cube of 280 × 565 spatial dimension. On the other hand, sometimes it is not necessary to keep all data corresponding to start and end of the spectra, thereby a spectral crop was conducted in order to reduce no essential data.

3.3.6. Smoothing spectra

In hyperspectral preprocessing, the use of smoothing methods to remove high-frequency noise signal on the reflectance spectra is regular; a quite common smoothing method used in remote sensing is the Savitzky-Golay filter [20], which is based on least-squares polynomial approach applied on the short steps of wavelengths. In this procedure, a window of 11 steps, with a polynomial degree 2, was used. **Figure 3** shows all spectra after preprocessing mentioned above. Each spectrum is the result of a region of interest (ROI) averaging a contiguous quadratic shape of nine pixels.

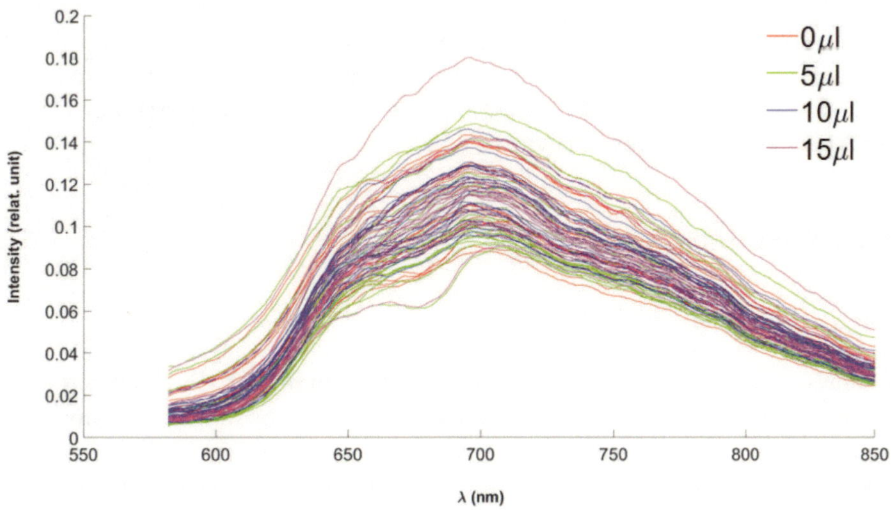

Figure 3. Spectra of 80 pixels extracted from 20 tomatoes; a ROI was selected in each contaminated zone.

3.3.7. *Obtaining modeling of spectral signatures*

On the other hand, inature a bunch of data distribution is frequently located as a Gaussian or normal distribution (as it is shown in **Figure 2**), so that this model relates directly the behavior of the datasets. Gaussian curve fitting is still investigated as an algorithm for detecting patterns in biological, social, and physical sciences [21].

In order to compute Gaussian models for each spectrum, MATLAB 2016a and curve fitting tool (cftool) were used. A total of 80 models were obtained; after several tests and errors, the best combination found for modeling was Gaussian polynomial model with five terms as the form of Eq. (2):

$$f(x) = a_1 e^{-\left(\frac{x-b_1}{c_1}\right)^2} + a_2 e^{-\left(\frac{x-b_2}{c_2}\right)^2} + a_3 e^{-\left(\frac{x-b_3}{c_3}\right)^2} + a_4 e^{-\left(\frac{x-b_4}{c_4}\right)^2} + a_5 e^{-\left(\frac{x-b_5}{c_5}\right)^2} \quad (2)$$

where $f(x)$ is the Gaussian model, x is the wavelength independent variable, and $a_1, a_2, a_3, a_4, a_5, b_1, b_2, b_3, b_4, b_5, c_1, c_2, c_3, c_4, c_5$ are the coefficients to be calculated.

3.3.8. *Computing area under curve and statistical analysis*

The calculation of all areas was carried out, by calculating the define integral (Eq. (3)). MATLAB 2016a provides an effective command called "quad" which numerically evaluates the integral, with an adaptive Simpson quadrature [22]:

$$A = \int_{wl_0}^{wl_1} f(x)dx \quad (3)$$

where A is the AUC; wl_0, wl_1 are the lower and upper limits, respectively, of wavelength; and $f(x)$ is a Gaussian model. Besides, the range between 582 and 850 nm was utilized, distributed into 482 bands.

After the areas under curves were obtained, a single factor analysis of variance (ANOVA) was performed in EXCEL 2016; the reason for this was to find ou if any relationship between the dosage amount (every 5 μL) and the decrease of the spectral signature response exists; a total of 20 tomatoes and 80 areas were analyzed.

4. Results and discussion

4.1. Gauss model results

Table 1 shows corresponding results of goodness of fit curve with Gaussians models. A low value in sum of squares error (SSE) is notorious, meaning that the model has a smaller random error component, since they are closer to zero [23]; as well as the coefficient of determination (R^2) has values higher than 0.9986; this proves high matching between the Gaussian model and the spectra responses to a certain dose. Besides, the other two parametric models for goodness of fit are adjusted R-sq and root-mean-square error (RMSE) which shows values higher than 0.99920 and less than 0.001, respectively.

http://dx.doi.org/10.5772/intechopen.73159

Another mathematical approach to know the good fitting of one predicted model is known as residuals, defined as the differences between the response of original data and the response to predicted model (Eq. (4)) with regard to recognizing if the model was

Dose (μL)	SSE	R^2	Adj R-sq	RMSE
0	0.00045494	0.99908	0.99905	0.00091
5	0.00026017	0.99939	0.99937	0.00071
10	0.0006778	0.99864	0.99860	0.00109
15	0.00037805	0.99922	0.99920	0.00085

Table 1. Results of fit models; every value is the media of fit result of each dose.

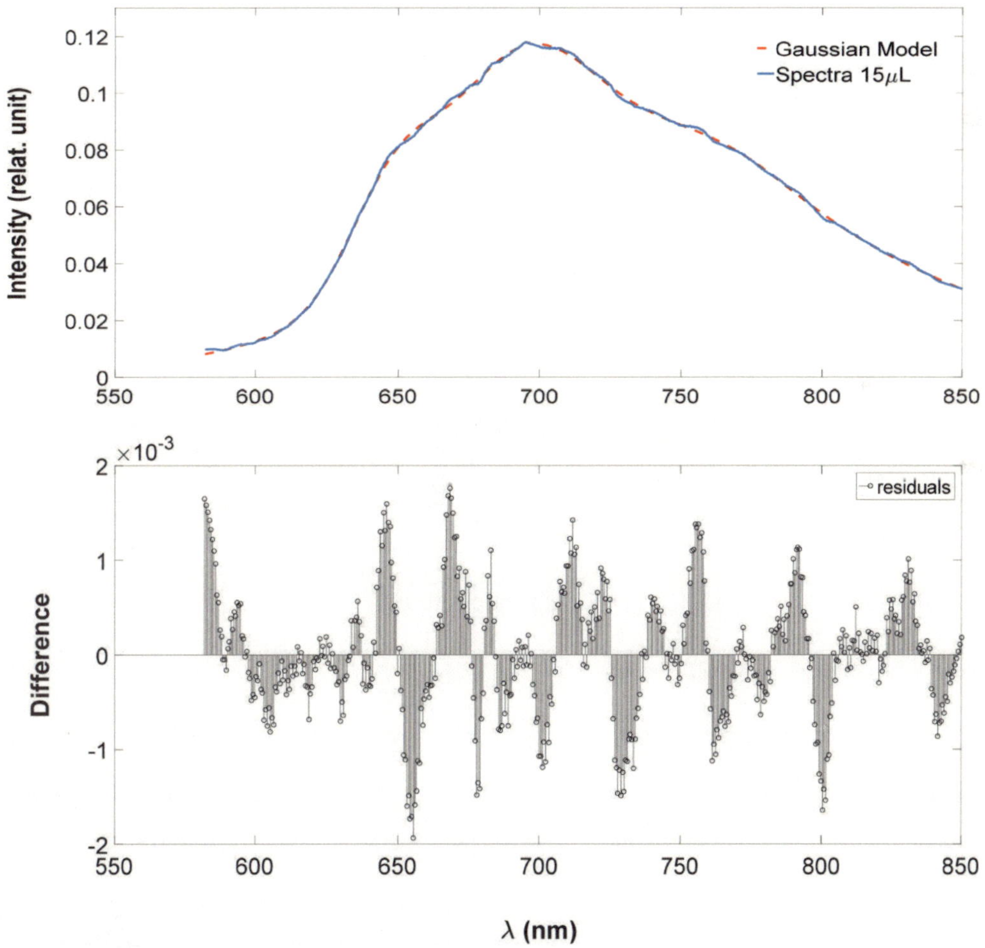

Figure 4. Upper graph shows an example of smoothed spectra (blue) and the predicted model (red); lower plot shows a random behavior on residuals, which means good prediction.

$$r = y - \widehat{y} \tag{4}$$

where r are the residuals, y are the spectra of contaminated zones, and $\widehat{y}$ is the predicted model. An example of residual response is shown in **Figure 4**. Whether the plot of residuals seems to behave in a random way, it means that the model fits the data well; otherwise, if residuals appear to behave in a systematic pattern, then it is a clear case of mismatch between data and model [24]. In this research, the whole 80 models showed random residuals.

4.2. Areas under curve and their analysis

Areas extracted from all spectral signatures are shown in **Figure 5**. The trend in this dataset seems to decrease with higher dose in most subsamples; the meaning of this is greater absorbance on the infected surface; as an exception, tomato surfaces 3, 4, 5, 6, 10, 11, 13, and 15 do not seem to have this behavior; one possible explanation is closely related with orientation and position at the time of hypercube acquisition, that is, little light saturation zones.

As a last step, the results of calculation for a single factor ANOVA are shown in **Table 2**. Because P-value <0.05 means that there is no significance between doses and spectra response, a similar methodology was conducted by [25].

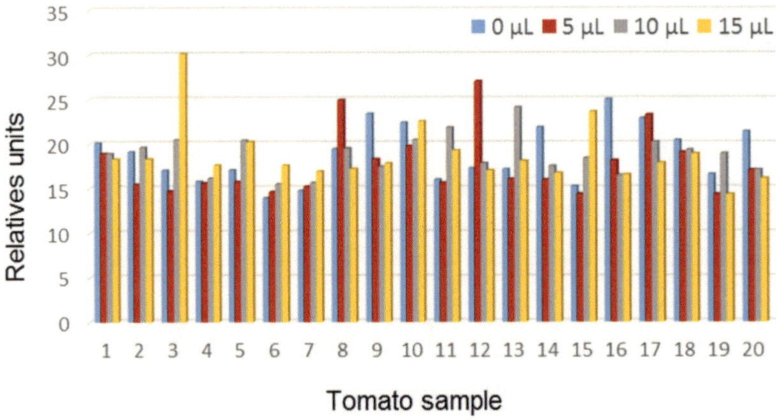

Figure 5. Total areas under curve of 20 tomatoes.

ANOVA						
Source of variation	**SS**	**df**	**MS**	**F**	**P-value**	**F crit**
Between groups	17.8996928	3	5.96656427	0.59978245	0.6171155	2.72494392
Within groups	756.038935	76	9.94788072			
Total	773.938627	79				

Table 2. Results of single factor ANOVA, taking AUC for data analysis.

5. Conclusion

Up to now, hyperspectral dataset analysis is carried out by different methodologies, algorithms, and techniques; in this research, we proposed to calculate AUC as an alternative for hypercubes; after AUC calculation, a single factor ANOVA would be enough for data analysis.

Despite results set down, it seems like visible range is not a good band for *S. typhimurium* detection. Secondly, sample orientation could improve results, since only a little inclination degree generated zone with high saturation because of the shiny nature of the tomato surface.

The novelty in this work was that there is little information related to the modeling of spectral signatures and their subsequent calculation of AUC as method to determine factors such as degree of contamination on fruits surface. Moreover, this methodology tries to quantify a spectral signature assigning it a value for understanding phenomenon that interacts with hyperspectral image systems. Future works could be related to improving AUC with different spectral responses in using variates fruit surfaces. Although there could be other variables to consider, which would affect the results as such, the scope of this work could be said to be a preliminar research.

Author details

Uziel Francisco Grajeda-González[1], Alejandro Isabel Luna-Maldonado[1]*, Humberto Rodriguez-Fuentes[1], Juan Antonio Vidales-Contreras[1], Ernesto Alonso Contreras Salazar[2] and Héctor Flores-Breceda[1]

*Address all correspondence to: alejandro.lunaml@uanl.edu.mx

1 Department of Agricultural and Food Engineering, Faculty of Agriculture, Autonomous University of Nuevo León, General Escobedo, Mexico

2 Department of Industrial Engineering, Culiacán Technological Institute, Culiacán, Mexico

References

[1] Wu D, Sun D-W. Advanced applications of hyperspectral imaging technology for food quality and safety analysis and assessment: A review—Part II: Applications. Innovative Food Science and Emerging Technologies. 2013;**19**:15-28. DOI: 10.1016/j.ifset.2013.04.016

[2] Bishop CM. Pattern Recognition and Machine Learning. New York: Springer-Verlag; 2006

[3] Ben-David S, Shalev-Shwartz S. Understanding Machine Learning: From Theory to Algorithms; 2014. DOI: 10.1017/CBO9781107298019

[4] Zhang YQ, Rajapakse JC. Machine Learning in Bioinformatics. John Wiley and Sons; 2009

[5] Brown AJ. Spectral curve fitting for automatic hyperspectral data analysis. IEEE Transactions on Geoscience and Remote Sensing. 2006;**44**(6):1601-1607. DOI: 10.1109/TGRS.2006.870435

[6] Hsu P-H. Feature extraction of hyperspectral images using wavelet and matching pursuit. ISPRS Journal of Photogrammetry and Remote Sensing. 2007;**62**(2):78-92. DOI: 10.1016/j.isprsjprs.2006.12.004

[7] Hsu PH, Tseng YH, Gong P. Dimension reduction of hyperspectral images for classification applications. Geographic Information Science. 2002;**8**(1):1-8. DOI: 10.1080/10824000209480567

[8] Gupta M, Jacobson N. Wavelet principal component analysis and its application to hyperspectral images. In: 2006 Int Conf Image Process. 2006. pp. 1585-1588. DOI: 10.1109/ICIP.2006.312611

[9] Gomez RB, Jazaeri A, Kafatos M. Wavelet-based hyperspectral multi-spectral image fusion. In: 2001 Geo-Spatial image and data exploitation II. International Society for Optics and Photonics. Vol. 4383. pp. 36-43

[10] Wang L, Gao K, Cheng X, Wang M, Miu X. A hyperspectral imagery anomaly detection algorithm based on gauss-Markov model. In: 2012 Fourth International Conference on Computational and Information Sciences. 2012. pp. 135-138. DOI: 10.1109/ICCIS.2012.21

[11] Malenovský Z, Ufer C, Lhotáková Z, et al. A new hyperspectral index for chlorophyll estimation of a forest canopy: Area under curve normalised to maximal band depth between 650-725 nm. EARSeL eProceedings. 2006;**5**:161-172. DOI: 10.5167/uzh-62112

[12] Yang H, Du Q, Su H, Sheng Y. An efficient method for supervised hyperspectral band selection. IEEE Geoscience and Remote Sensing Letters. 2011;**8**(1):138-142. DOI: 10.1109/LGRS.2010.2053516

[13] Chang CI, Wang S. Constrained band selection for hyperspectral imagery. IEEE Transactions on Geoscience and Remote Sensing. 2006;**44**(6):1575-1585. DOI: 10.1109/TGRS.2006.864389

[14] Li S, Zhu Y, Wan D, Feng J. Hyperspectral band selection from the spectral similarity perspective. In: International Geoscience and Remote Sensing Symposium (IGARSS); 2013. pp. 410-413. DOI: 10.1109/IGARSS.2013.6721179

[15] Peng Y, Lu R. Analysis of spatially resolved hyperspectral scattering images for assessing apple fruit firmness and soluble solids content. Postharvest Biology and Technology. 2008;**48**(1):52-62. DOI: 10.1016/j.postharvbio.2007.09.019

[16] Yoon S-C, Shin T-S, Lawrence KC, Heitschmidt GW, Park B, Gamble GR. Hyperspectral imaging using RGB color for foodborne pathogen detection. Journal of Electronic Imaging. 2015;**24**(4):43008. DOI: 10.1117/1.JEI.24.4.043008

[17] Cito F, Baldinelli F, Calistri P, et al. Outbreak of unusual Salmonella enterica serovar Typhimurium monophasic variant 1, 4[5], 12:I:-, Italy, June 2013 to September 2014. Eurosurveillance. 2016;**21**(15):1-10. DOI: 10.2807/1560-7917.ES.2016.21.15.30194

[18] Ribera LA, Palma MA, Paggi M, Knutson R, Masabni JG, Anciso J. Economic analysis of food safety compliance costs and foodborne illness outbreaks in the United States. HortTechnology. 2012;**22**(2):150-156

[19] James B, Geladi P. Hyperspectral NIR image regression part II: Dataset preprocessing diagnostics. Journal of Chemometrics. 2006;**20**(3-4):106-109. DOI: 10.1002/cem

[20] Schmidt KS, Skidmore AK. Smoothing vegetation spectra with wavelets. International Journal of Remote Sensing. 2004;**25**(6):1167-1184. DOI: 10.1080/0143116031000115085

[21] Gauch HGJ, Chase GB. Fitting the Gaussian curve to ecological data. Ecology. 1974;**55**(6): 1377-1381. DOI: 10.2307/1935465

[22] MATLAB. Version 9.10.0 (R2016a). Natick, Massachusetts: The MathWorks Inc.; 2016

[23] Steinley D. Validating clusters with the lower bound for sum-of-squares error. Psychometrika. 2007;**72**(1):93-106. DOI: 10.1007/s11336-003-1272-1

[24] NIST/SEMATECH. e-Handbook of Statistical Methods. 2012. http://www.itl.nist.gov/div898/handbook/

[25] Pruessner JC, Kirschbaum C, Meinlschmid G, Hellhammer DH. Two formulas for computation of the area under the curve represent measures of total hormone concentration versus time-dependent change. Psychoneuroendocrinology. 2003;**28**(7):916-931. DOI: 10.1016/S0306-4530(02)00108-7

Section 3

Hyperspectral Imaging in Agriculture

Chapter 6

Sequential Classification of Hyperspectral Images

Min Zhao and Jie Chen

Additional information is available at the end of the chapter

http://dx.doi.org/10.5772/intechopen.73160

Abstract

Hyperspectral imaging has become increasingly popular in applications such as agriculture, food, and environment. Rich spectral information of hyperspectral images leads to new possibilities and new challenges in data processing. In this chapter, we consider the hyperspectral classification problems in consideration of sequential data collection, which is a frequent setting in industrial pushboom imaging systems. We present related techniques including data normalization, dimension reduction, classification, and spatial information integration and the way to accommodate these techniques to the context of sequential data collecting and processing. The propose scheme is validated with real data collected in our laboratory. The methodology of result assessment is also presented.

Keywords: hyperspectral sorting, sequential hyperspectral data processing, spatial-spectral information, hyperspectral classification

1. Introduction

Hyperspectral imaging is a continuously growing area and has received considerable attention in the last decade. Hyperspectral data provide a wide spectral range, coupled with a high-spectral resolution. These characteristics are suitable for detection and classification of surfaces and chemical elements in the observed images. Rich information in spectral dimension provides solutions to many problems that cannot be solved by traditional RGB imaging or multispectral imaging.

Applications include land use analysis, pollution monitoring, wide-area reconnaissance, and field surveillance, to cite a few. Typical cases related to food quality, agriculture, and environment include as follows:

1. Food safety plays an important role in our daily life. We often use a combination of appearance, hand-feel, and smell of the product to make a judgment of the quality of fruits

or vegetables. But it is not enough to judge if there are abnormalities, deformations, or even visible defects in the fruit or vegetable. Awareness about food safety has exemplified the requirement for a rapid and accurate hyperspectral detection system [1].

2. Precision agriculture is a farming management concept based on observing, measuring, and inter and intrafield variability in crops. Precision agriculture using hyperspectral remote sensing is acquired and processed to derive maps of crop biophysical parameters, to measure the amount of plant cover, and to distinguish between crops and weeds [2].

3. Due to the pressures of over consumption, population, and technology, the biophysical environment is being degraded, sometimes permanently. Many of the earth's resources are on the verge of exhaustion because they are influenced by human impacts across many countries [3]. Many attempts are made to prevent damage or manage the impacts of human activity on natural resources. Hyperspectral classification used in resource recovery can make it rapid and efficient.

One of the most important tasks of hyperspectral image processing is image classification. Rich spectral information of hyperspectral image provides the possibility to classify materials that are difficult to be distinguished by other imagery techniques. In the past decades, different kinds of hyperspectral classification methods have been proposed [4–9]. However, the existed methods may not be suitable for a real-time material sorting system. Pushboom imaging systems are frequently used in industry sorting, such a system collects columns of an image one after another in a sequential manner (see **Figure 1**). It is thus necessary to design a framework for online classification tasks and accommodate convectional algorithms to the sequential processing setting.

In this chapter, we present a scheme of sequential classification for hyperspectral sorting systems. This scheme can be used in various fields, such as measuring food quality and resource recovery. We present the main techniques in this sorting and processing, including data normalization, dimension reduction, classification, and spatial information integration and the way to accommodate these techniques to the context of sequential data collecting and processing.

The rest of this chapter is organized as follows. In Section 2, we propose the main steps of sequential hyperspectral classification processing system. In Section 3, detailed methods are

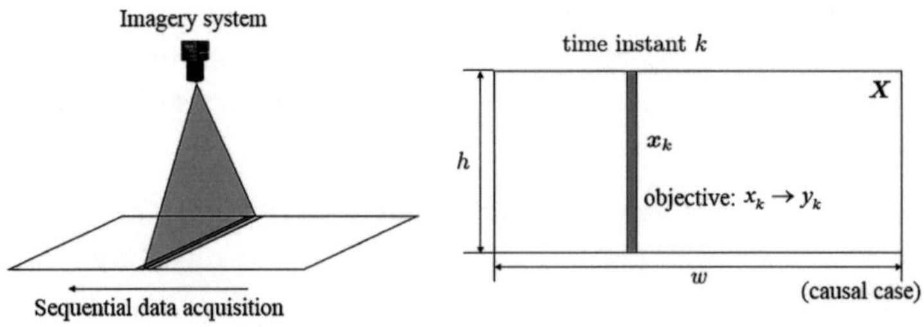

Figure 1. Sequential hyperspectral data collecting and processing by a pushboom system. Hyperspectral camera captures data x_k at time instant k, which is one of the sequential columns of the entire image, y_k is the result after processing (classification label in this case).

presented for sequential hyperspectral image processing and sorting. Experiment results are then discussed in Section 4. Section 5 concludes the chapter.

2. System overview

Before proceeding to elaborate the proposed sequential hyperspectral image classification method, we first present the notation and the data model used in this work. We consider that the hyperspectral image under study has h pixels in column and w pixels in row, where h is a fixed size that is determined by the spatial resolution of the camera, and w actually increases toward infinity with the moving of the pushboom system. Each pixel consists of a reflectance vector consisting of p contiguous spectral bands. Then, let

- $N = h \times w$ be the total number of pixels.
- $X = \{X_1, X_2, X_3, \cdots, X_p\}^T$ be the $(p \times N)$ hyperspectral images.
- $X_k(i,j)$ represents a pixel, where the subscript k denotes the index of the spectral band, i and j represent the location of pixel in the spatial domain.

The data collecting and processing of a real-time hyperspectral sorting system consist of the following major steps.

1. Sequential image acquisition.
2. Data preprocessing.
3. Material classification.

The hyperspectral data used in this work set are collected by the system of GaiaField in our laboratory. The parameters of the used system are provided in **Table 1**. Our online processing is based on windowed columns. After collecting each column, we use this column with several previous ones to form a window and perform data processing steps within this window. Black-white normalization is used for basic data normalization. Techniques of PCA and hyperspectral decorrelation of fuzzy sets are used for dimension reduction [10]. Typical techniques such as GML and SVM are presented for material classification. Considering the positive effect of spatial information on processing results [11], we also propose to integrate

Equipment type	GaiaField and GaiaSorter
Moving speed of loading platform	4.1 cm/s
Spectral resolution	128
Spatial resolution	650 × 348
Distance between lens and samples	24 cm
Exposure time	3 ms

Table 1. Equipment parameters.

spatial dimension and spectral dimension to achieve an enhanced classification accuracy. Finally, classification accuracy is characterized by metrics such as confusion matrix and κ coefficient. Details of the used techniques and results will be provided later.

3. Processing methods

3.1. Data preprocessing

Data preprocessing steps include basic data normalization and spectral decorrelation. They are performed one after another as described later.

3.1.1. Basic data normalization

An important preprocessing is the so-called black-white calibration. This calibration is carried out by recording an image for black and another for white, as described below, to remove the effect of dark current of the camera sensor and avoid the uneven light intensity of each band. At an offline phase, the black image (*B*) is acquired by turning off the light source and covering the camera lens with its cap. The white image (*W*) is acquired by adopting a standard white ceramic tile under the same conditions as the raw image. Then, image correction is performed by [12],

$$I = \frac{I_0 - B}{W - B} \tag{1}$$

where I is the hyperspectral image after normalization, I_0 is the original hyperspectral image that is captured in our laboratory, B is the black reference image, and W is the white reference image.

3.1.2. Data dimension reduction

The high-spectral resolution of hyperspectral data enables us to classify materials that are undistinguishable with conventional methods. However, a large number of spectral channels result in difficulties in processing in terms of classifier training (Hughes phenomenon) and computational burdens. Data dimension reduction can be performed due to the above facts and existence of information redundancies across bands.

3.1.2.1. PCA

PCA is one of the most popular methods for data dimension reduction. PCA computes a linear transformation for high-dimensional input vectors, and this transformation maps the data into a low-dimensional orthogonal subspace. For simplicity, we assume that the data samples have zero mean. Otherwise, we can centralize the data by subtracting the mean

$$Y = X - E(X) \tag{2}$$

The principle analysis is based on the eigenstructure of the data. We therefore calculate the covariance matrix of Y and perform the eigendecomposition on this matrix. The ith eigenvector of matrix Y is denoted by a_i with associated eigenvalue denoted by λ_i.

To reduce the dimension of data, we select an appropriate number of eigenvectors a_i corresponding to the value of eigenvalues λ_i from large to small, to form the representation coefficient matrix A [13].

$$Z = A^T X \tag{3}$$

where Z is the hyperspectral image data after decorrelation.

3.1.2.2. *Fuzzy sets*

Using fuzzy sets to decorrelate the hyperspectral data is based on a priori knowledge that the adjacent wavelengths of the spectrum are more correlated than the distant pairs, as the spectral information varies smoothly and successively. We consider sampling spectral characteristics by a group of adjacent spectral bands, which can be obtained by dividing the spectra in separate groups to attain the desired spectral selectivity. We propose separating the hyperspectral data into a number of M fuzzy groups where each group covers a range of wavelengths [14]. The contribution of each wavelength is modeled by a membership function $Mf_i(\lambda)$. We use a triangular function as the membership function, shown in **Figure 2**.

$$Mf_i = \begin{cases} 1 - \dfrac{\lambda - \lambda_i}{D}, & \lambda_i - D < \lambda < \lambda_i + D \\ 0 & \text{otherwise} \end{cases} \tag{4}$$

where λ_i is the central wavelength value of the fuzzy set i, and D is the distance of central wavelengths of two adjacent fuzzy sets.

The spectral wavelengths of all points have different membership degrees in different fuzzy sets. Each wavelength has different degree of membership in two adjacent fuzzy sets, while the membership degree in the remaining fuzzy sets is 0 (**Figure 3**).

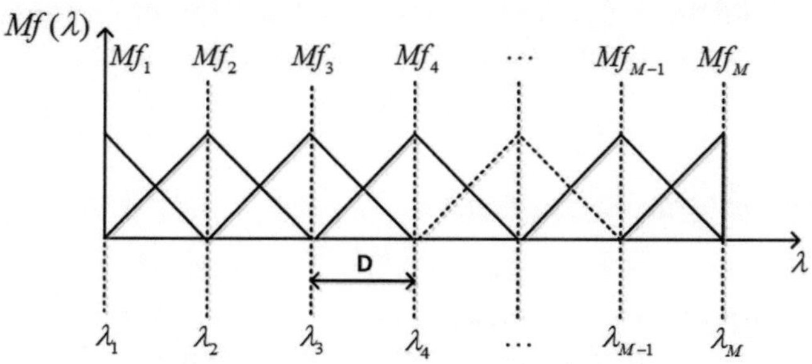

Figure 2. Triangular function.

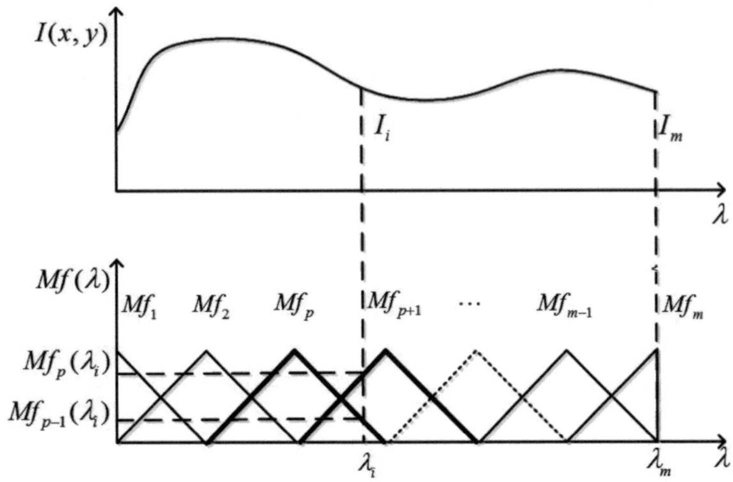

Figure 3. Triangular function weighted process.

The energy of each fuzzy set is calculated by weighting the intensity of each spectrum element using membership functions associated with each fuzzy set, i.e.,

$$X_i = \int_{\lambda=0}^{\lambda=K} Mf_i \cdot L(\lambda) d\lambda \tag{5}$$

where X_i is the energy of each fuzzy set, and $L(\lambda)$ is the intensity of each spectrum element.

Based on the energy values of each fuzzy set, we can obtain useful information about the spectral characteristics. In this way, each hyperspectral image pixel can be defined by a vector containing the energy values of the M fuzzy sets as

$$X = \{X_1, X_2, \cdots, X_M\}^T \tag{6}$$

3.2. Material classification

In this section, we present the algorithm to classifier/sort the captured data using features (data of reduced dimension) extracted by PCA or fuzzy set method. We first review these two popular classification methods in a general manner. Then we introduce how to incorporate spatial information into the classification. Finally, sequential processing with window-based method will be discussed.

3.2.1. Gaussian maximum likelihood classification

Spectra of distinct material of hyperspectral data form data clusters in a space with the dimension of the feature, and we assume that the data features of each material approximately follow a multivariate normal distribution. To be specific, data features of a material i and the p dimension probability density function in form of:

http://dx.doi.org/10.5772/intechopen.73160

$$p\left(X|\omega_i\right) = \left(\frac{1}{2\pi}\right)^{\frac{p}{2}} |\Sigma_i|^{-\frac{1}{2}} \exp\left[-\frac{1}{2}\left(X-\mu_i\right)^T \Sigma_i^{-1}\left(X-\mu_i\right)\right] \tag{7}$$

where μ and Σ are the mean vector and the covariance matrix, respectively. i denotes the label of class [15]. Each pixel in the hyperspectral image is labeled as the class that achieves maximum probability.

3.2.2. *Support vector machine*

SVM is one of the most effectively and widely used methods in statistical learning. SVM aims to find the best tradeoff of model complexity and learning ability with limited sample information. SVM can effectively solve the Hughes phenomenon caused by insufficient samples in hyperspectral classification.

The goal of training algorithm is to design an optimal hyperplane. The training principle of SVM is to find a linear optimal separating hyperplane [16]. Let x be the input pixel vectors satisfying

$$g(x) = \omega^T x + \omega_0 = 0 \tag{8}$$

This method constructs a hyperplane that maximizes the margins between classes, specified by a (usually small) subset of the data that define the position of the separator. These points are referred to as the support vectors [17]. The decision function is as follow:

$$f(x) = \text{sgn}\left[\omega^T x + b\right] = \text{sgn}\left(\sum_{i=1}^{l} \alpha_i y_i {x_i}^T x + b\right) \tag{9}$$

where α_i is the ith Lagrange coefficient, y_i is the corresponding classification label, x_i is the ith support vector, x is the input pixel vector, N is the number of support vector, and b is the decision offset coefficient. For two-class hyperspectral classification, $f(x)$ takes value of either 1 or 0, suggesting the class that the current pixel belongs. For multiclass classification, we can use one versus one, one versus rest, hierarchical support vector machine or other strategy to obtain the multiclass label.

Sometimes, data cannot be separated by a linear classifier. Therefore, kernel methods are used to map data from the original input space to a higher dimension space. Thanks to the kernel trick, we only need to know the form of the inner product in that space instead of using the explicit map [16]. Popular kernel functions include as follows:

Linear kernel:

$$K\left(x, x_j\right) = x \cdot x_j. \tag{10}$$

Polynomial kernel:

$$K(x, x_i) = \left[\left(x^T x_i\right) + 1\right]^q, \quad q > 0 \tag{11}$$

where q is the polynomial order.

Radial basis function kernel:

$$K(x, x_i) = \exp\left(-\frac{-\|x - x_i\|^2}{\sigma^2}\right) \tag{12}$$

where σ^2 is kernel bandwidth.

Sigmoid kernel:

$$K(x, x_i) = \tanh\left[v\left(x^T x_i\right) + c\right] \tag{13}$$

for appropriate values of v and c, so that Mercer's conditions are satisfied [16].

3.2.3. Incorporating spatial information

Conventionally, hyperspectral data classification algorithms are proposed based on spectroscopic viewpoint, and they ignore the spatial information that embeds in neighboring pixels [18]. Integration of spatial and spectral information may improve the processing performance. We propose to combine spatial dimension and spectral dimension information to improve the classification accuracy. The proposed method investigates the spatial information based on the connection component labeling in the following. We generate the mean image by averaging data after dimension reduction over spectra bands. A component labeling algorithms then applied to the binarized mean image. In our system, if an object is marked by connected component labeling and over 60% pixels are labeled as a class, we consider that all pixels within this connected region actually belong to the associated material. The classification accuracy will be improved using this strategy.

3.2.4. Sequential processing

We use a sliding window to assemble the acquired hyperspectral data, whose columns are collected sequentially one after another. The use of a sliding window facilitates to incorporate the spatial information in processing. The width of the sliding window should be determined by considering the data acquisition rate, data processing speed, and spatial correlation of the observed scenario. In our system, the width of the sliding window (L) is set to 15. Our hyperspectral images are captured by a pushboom system where columns of images are collected sequentially one after another. After collecting each column, we use this column with several previous ones to form a sliding window and perform data processing steps within this window. Let L be the width of the sliding window, and we set $L = 15$ in our experimental (**Figure 4**).

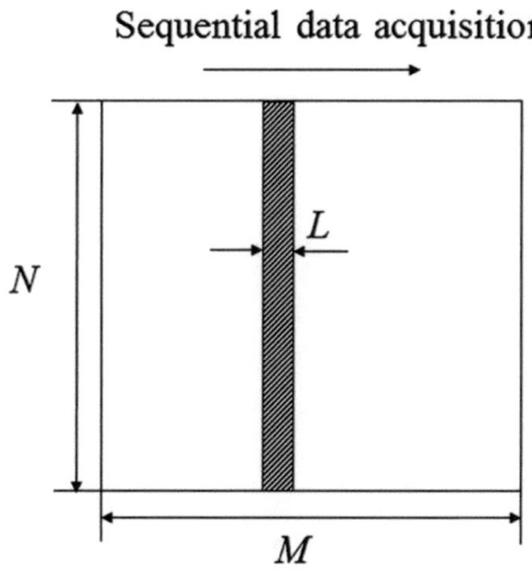

Figure 4. Sliding window.

4. Experimental results

We collect the hyperspectral data with our pushboom system of Gaia. The images are acquired in the 400–1000 nm wavelength range, with a spectral resolution of 7 nm, for a total of 128 wavelengths ($p = 128$). Their image resolutions h and w are 650 and 348 (650×348), respectively. The hyperspectral data include four kinds of fruits: tomato, jujube, lemon, and orange. In this study, we use a sliding window of size 15 for online processing of data. Twenty-three sequential hyperspectral images are extracted for classification. The datasets captured are divided into training and testing sets, where 300 pixels of each material are used for training and 30,603 pixels are used for testing.

After data preprocessing, we select 300 pixels of each material from the training set as sample points to form a hyperspectral image. The pixels of the image are converted into row vectors by row or column to form a two-dimensional matrix, which is used for data reduction. The operation of the test set is the same as that of the training set.

After the PCA transformation, the eigenvalue distribution is shown in **Figure 5**. This scree plot shows that the first eight factors explain most of the variability. The remaining factors explain a very small proportion of the variability and are likely unimportant. We select the principal component, which takes 99% of the eigenvalues, as the data after dimensionality reduction. For fuzzy-set data reduction, we fold 128 bands with a triangular window of length 32, and then we sample the data using at each 16 points, so that the data dimension also reduces to 8. We use eight-connected component labeling method to remove the background of data after dimension reduction.

We then study the classification results of GML principle and SVM. We classify the data obtained from dimension reduction and background material removal (see **Figures 6** and **7**). The result of classification with spatial information (connected region labeling) is shown in **Figure 8**.

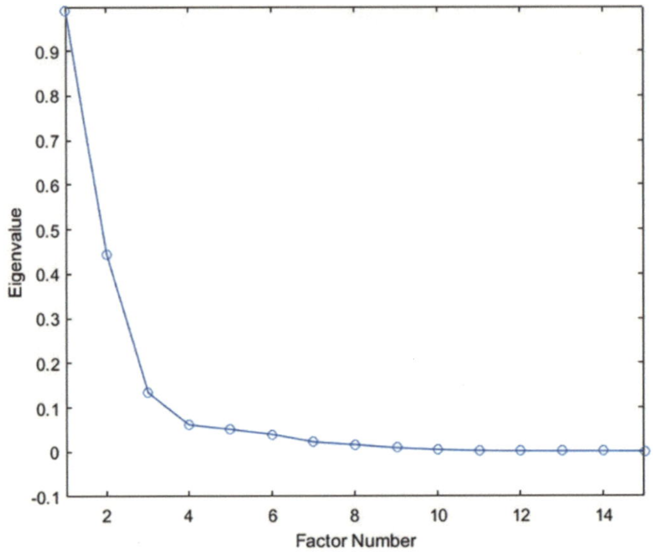

Figure 5. Eigenvalue distribution.

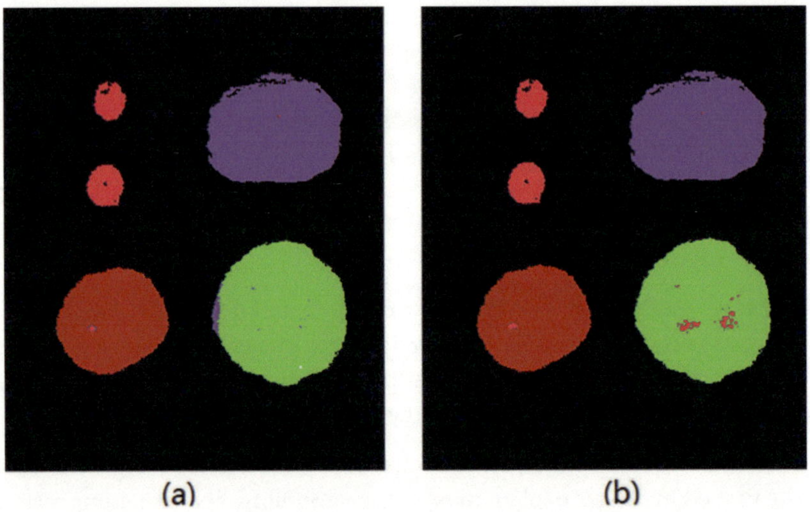

Figure 6. (a) GML classification with PCA and (b) GML classification with fuzzy sets.

4.1. Performance evaluation of results

4.1.1. Confusion matrix

A confusion matrix is a table that is often used to describe the performance of a classifier on a set of test data for which the true values are known. It compares the classification result with the reference image, and we need to determine the labels of each point in the reference image in the classified image. The confusion matrixes of our experiment are shown in **Table 2**.

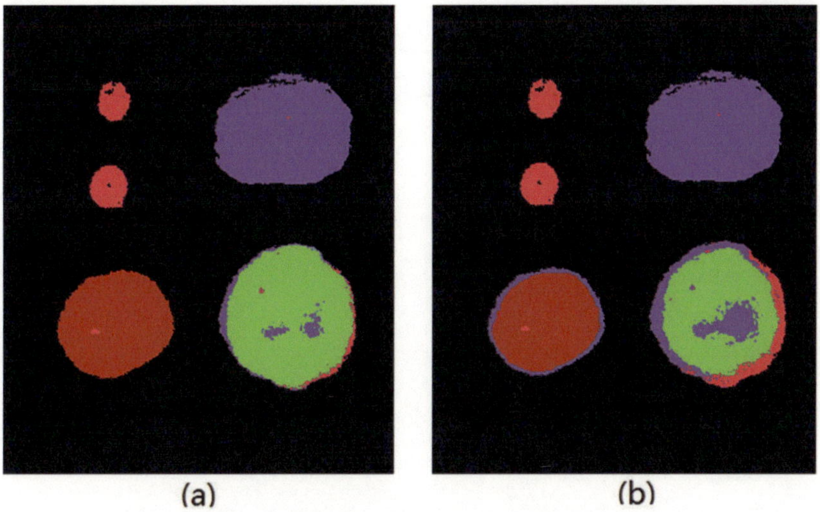

Figure 7. (a) SVM classification with PCA and (b) SVM classification with fuzzy sets.

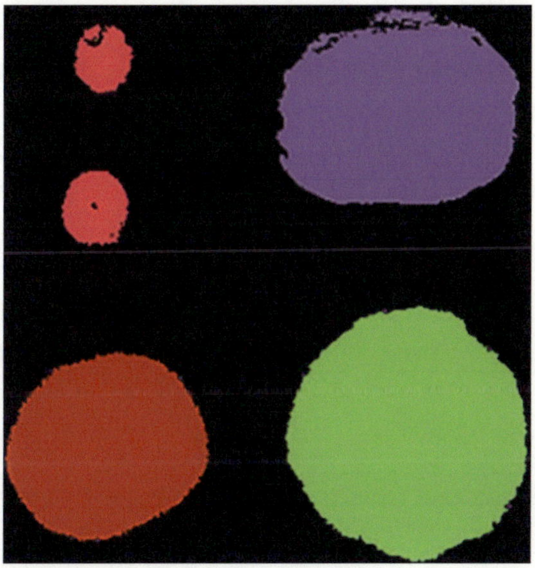

Figure 8. Classification with spatial information (connected region labeling) achieves almost 100% accuracy.

$$A = \begin{bmatrix} m_{11} & m_{12} & \cdots & m_{1k} \\ m_{21} & m_{22} & \cdots & m_{2k} \\ \vdots & \vdots & & \vdots \\ m_{k1} & m_{k2} & \cdots & m_{kk} \end{bmatrix} \tag{14}$$

where m_{ij} shows pixels should belong to class i, which is wrongly assigned to class j, and k is the class number of the classification results (**Figures 9** and **10**).

Class		Actual class				
		Chinese date	Lemon	Orange	Tomato	Row sum
(1) GML classification with PCA						
Predict class	Chinese date	1745	0	0	0	1745
	Lemon	4	9618	0	0	9622
	Orange	0	183	11,654	0	11,837
	Tomato	26	14	0	7359	7399
	Column sum	1775	9815	11,654	7359	30,603
(2) GML classification with fuzzy sets						
Predict class	Chinese date	1745	0	0	0	1745
	Lemon	3	9619	0	0	9622
	Orange	258	2	11,577	0	11,837
	Tomato	36	7	0	7356	7399
	Column sum	2042	9628	11,577	7356	30,603
(3) SVM classification with PCA						
Predict class	Chinese date	1744	1	0	0	1745
	Lemon	6	9616	0	0	9622
	Orange	466	1089	10,282	0	11,837
	Tomato	29	0	0	7370	7399
	Column sum	2245	10,706	10,282	7370	30,603
(4) SVM classification with fuzzy sets						
Predict class	Chinese date	1738	0	0	7	1745
	Lemon	5	9617	0	0	9622
	Orange	1489	2826	7522	0	11,837
	Tomato	28	1144	0	6227	7399
	Column sum	3260	13,587	7522	6234	30,603

Table 2. Confusion matrix of classification results: (1) GML classification with PCA, (2) GML classification with fuzzy sets, (3) SVM classification with PCA, and (4) SVM classification with fuzzy sets.

4.1.2. *κ coefficient*

κ can reflect the classification error of the whole image and solve the problem that the classification accuracy depends too much on the number of classes and the number of samples. κ is performed by adopting the following equation:

$$\kappa = \frac{N\sum_{i=0}^{k} m_{i,i} - \sum_{i=0}^{k} m_{i+} \times m_{+i}}{N^2 - \sum_{i=0}^{k} m_{i+} \times m_{+i}} \tag{15}$$

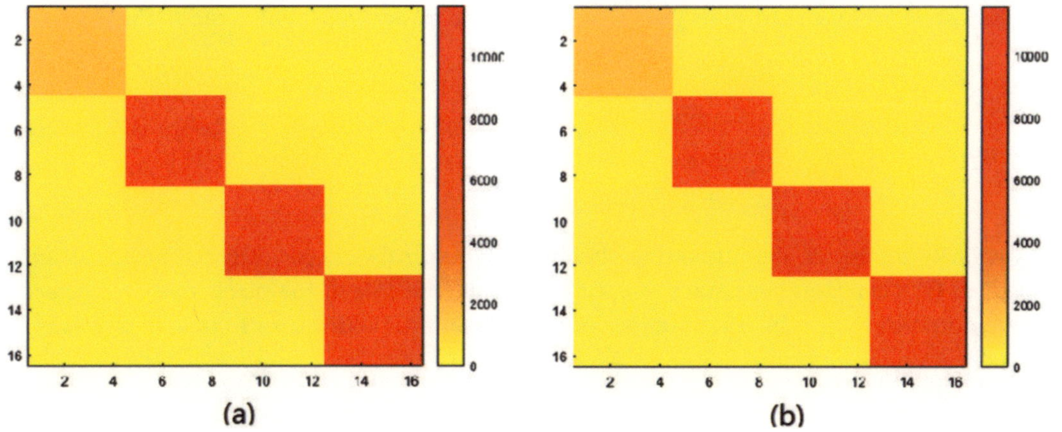

Figure 9. (a) Confusion matrix of GML classification with PCA and (b) confusion matrix of GML classification with fuzzy sets.

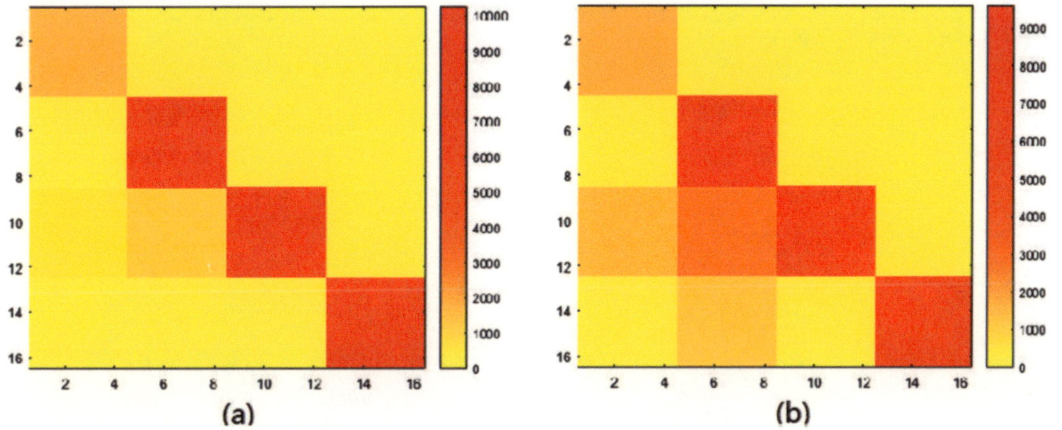

Figure 10. (a) Confusion matrix of SVM classification with fuzzy sets and (b) confusion matrix of SVM classification with fuzzy sets.

where m_{i+} is the sum of the line i in the confusion matrix, and m_{+i} is the sum of the column i in the confusion matrix.

κ of GML based on PCA dimensionality reduction is 98.93%, and κ of SVM is 92.55%. κ of GML based on fuzzy-set reduction technique is 98.56%, and κ of SVM is 74.69%. From the results of κ, we can see that the classification based on PCA is better than fuzzy sets, GML is better than SVM, and GML based on PCA is the best method for sequential classification of hyperspectral images.

4.1.3. Other metrics

Other metrics include classification accuracy, product's accuracy (PA), and omission errors (OEs).

Classification accuracy indicates the correct rate of the classifier, as illustrated in Eq. (16).

$$A_i = \frac{\sum_{i=0}^{k} m_{i,i}}{\sum_{i=0}^{k} m_{+i}} \tag{16}$$

PA is used to indicate the rate of the classification result that is correctly classified, as illustrated in Eq. (17). User's accuracy is used to indicate the rate of the pixels that are correctly divided into class I to the total number of pixels that are divided into I classes, as shown in Eq. (18).

$$PA_i = \frac{m_{i,i}}{m_{+i}} \tag{17}$$

Other metrics	Chinese date	Lemon	Orange	Tomato
(1) GML classification with PCA				
PA_i	0.9831	0.9799	1	1
OE_i	0.0169	0.0201	0	0
UA_i	1	0.9996	0.9845	0.9946
CE_i	0	0.0004	0.0155	0.054
(2) GML classification with fuzzy sets				
PA_i	0.8546	0.9991	1	1
OE_i	0.1454	0.0009	0	0
UA_i	1	0.9997	0.9780	0.9941
CE_i	0	0.0003	0.0220	0.0059
(3) SVM classification with PCA				
PA_i	0.7768	0.8982	1	1
OE_i	0.2232	0.1018	0	0
UA_i	1	0.9994	0.8686	0.9961
CE_i	0	0.0006	0.1314	0.0039
(4) SVM classification with fuzzy sets				
PA_i	0.5331	0.7078	1	0.9989
OE_i	0.4669	0.2922	0	0.0011
UA_i	0.9960	0.9995	0.6355	0.8416
CE_i	0.0040	0.0005	0.3645	0.1584

Table 3. Other metrics: (1) GML classification with PCA, (2) GML classification with fuzzy sets, (3) SVM classification with PCA, and (4) SVM classification with fuzzy sets.

$$UA_i = \frac{m_{i,i}}{m_{i+}} \tag{18}$$

OEs represent the number of pixels in class I that is incorrectly assigned to other class, as shown in Eq. (19). Commission errors (CEs) indicate the percentage of other class pixels that are incorrectly divided into class I, as illustrated in Eq. (20).

$$OE_i = 1 - \frac{m_{i,i}}{m_{+i}} \tag{19}$$

$$CE_i = 1 - \frac{m_{i,i}}{m_{i+}} \tag{20}$$

Classification accuracy of GML based on PCA dimensionality reduction is 99.26%, and classification accuracy of SVM is 94.80%. Classification accuracy of GML based on fuzzy-set reduction technique is 99.00%, and classification accuracy of SVM is 82.03%. From this evaluation and **Table 3**, GML based on PCA dimensionality reduction is the proposed solution for sequential classification of hyperspectral images.

5. Conclusion

The major objective of this chapter is to build a sequential hyperspectral classification method for an industrial material sorting system. We propose hyperspectral images captured by a pushboom system where columns of images are collected sequentially one after another to get sequential hyperspectral images. PCA and fuzzy sets are used for data decorrelation. We study the GML and SVM classification with the data obtained from dimension reduction and background material removal and carry out the performance analysis. The results show that the accuracy rate of GML based on PCA dimensionality reduction is 99.26%, and the accuracy rate of SVM is 94.80%. The accuracy of GML based on fuzzy-set reduction technique is 99.00%, and the accuracy rate of SVM is 82.03%. After combing the spatial and spectral information, the accuracy of classification of hyperspectral images can be 100%.

The designed framework shows several advantages in terms of processing speed, efficiency, and accuracy. It may play an important role in industrial material sorting for agricultural products, food, and industrial waste sorting.

Acknowledgements

This work was supported in part by National Natural Science Foundation of China under grant 61671382 and in part by National Natural Science Foundation of Shenzhen under grant JCYJ2017030155315873.

Author details

Min Zhao[1,2] and Jie Chen[1,2]*

*Address all correspondence to: dr.jie.chen@ieee.org

1 Centre of Intelligent Acoustics and Immersive Communications School of Marine Science and Technology, Northwestern Polytechnical University, China

2 Research & Development Institute of Northwestern Polytechnical University in Shenzhen, China

References

[1] Loutfi A, Coradeschi S, Mani GK, Shankar P, Rayappan JBB. Electronic noses for food quality: A review. Journal of Food Engineering. 2015;**144**:103-111

[2] Cetin, H and Pafford, JT and Mueller, TG. Precision agriculture using hyperspectral remote sensing and GIS. Recent Advances in Space Technologies, 2005. RAST 2005. Proceedings of 2nd International Conference on. 2005. pp. 70-77

[3] Bonifazi G, Serranti S, Bonoli A, Dall'Ara A. Innovative recognition-sorting procedures applied to solid waste: The hyperspectral approach. WIT Transactions on Ecology and the Environment. 2009;**120**:885-894

[4] Harsanyi JC, Chang C-I. Hyperspectral image classification and dimensionality reduction: An orthogonal subspace projection approach. IEEE Transactions on geoscience and remote sensing. 1994;**32**:779-785

[5] Ye Z, Prasad S, Li W, Fowler JE, He M. Classification based on 3-D DWT and decision fusion for hyperspectral image analysis. IEEE Geoscience and Remote Sensing Letters. 2014;**11**:173-177

[6] Camps-Valls G, Gomez-Chova L, Munoz-Mari J, Vila-Frances J, Calpe-Maravilla J. Composite kernels for hyperspectral image classification. IEEE Geoscience and Remote Sensing Letters. 2006;**3**:93-97

[7] Melgani F, Bruzzone L. Classification of hyperspectral remotesensing. IEEE Transactions on Geoscience and Remote Sensing. 2004;**42**:1778-1790

[8] Palmason JA, Benediktsson JA, Sveinsson JR, Chanussot J. Fusion of morphological and spectral information for classification of hyperspectral urban remote sensing data. In: Proc. IGARSS. 2006. pp. 2506–2509

[9] Chi M, Bruzzone L. Semisupervised classification of hyperspectral. IEEE Transactions on Geoscience and Remote Sensing. 2007;**45**:1870-1880

[10] Artzai, Ghita O, Bereciartua A, Echazarra J, Whelan PF, Iriondo PM. Real-time hyperspectral processing for automatic nonferrous material sorting. Journal of Electronic Imaging. 2012;**21**: 013018-1

[11] Wahab DA et al. Development of a prototype automated sorting system for plastic recycling. American Journal of Applied Sciences. 2006;**3**(7):1924-1928

[12] Serranti S, Palmieri R, Bonifazi G. Hyperspectral imaging applied to demolition waste recycling: Innovative approach for product. Journal of Electronic Imaging. 2015;**24**(4): 043003

[13] Wold S, Esbensen K, Geladi P. Principal component analysis. Chemometrics and Intelligent Laboratory Systems. 1987;**2**:37-52

[14] Picón A et al. Fuzzy spectral and spatial feature integration for classification of nonferrous materials in hyperspectral data. IEEE Transactions on Industrial Informatics. 2009; **5**(4):483-494

[15] Bishop CM. Pattern Recognition and Machine Learning. New York: Springer; 2006. ISBN-10:0-387-31073-8

[16] Demir B, Erturk S. More sparsity in hyperspectral SVM classification using unsupervised pre-segmentation in the training phase. Recent Advances in Space Technologies, 2007. RAST'07. 3rd International Conference on. 2007;271-274

[17] Wu Y. Approximate computing of remotely sensed data: SVM hyperspectral image classification as a case study. IEEE Journal of a Selected Topics in Applied Earth Observations and Remote Sensing. 2016;**9**:5806-5815

[18] Stockman H, Gevers T. Detection and classification of hyper-spectral edges. In: British Machine Vision Conf. (BMVC); Nottingham, UK. 1999. pp. 643-651

Section 4

Hyperspectral Imaging in Environment

Chapter 7

Soil Contamination Mapping with Hyperspectral Imagery: Pre-Dnieper Chemical Plant (Ukraine) Case Study

Sergey A. Stankevich, Mykola M. Kharytonov,
Anna A. Kozlova, Vadym Yu. Korovin,
Mykhailo O. Svidenyuk and Alexander M. Valyaev

Additional information is available at the end of the chapter

http://dx.doi.org/10.5772/intechopen.72601

Abstract

Radioactive contamination of soils is an issue of severe importance for Ukraine remaining with a significant post-Soviet baggage of not settled problems regarding radioactive waste. Regular radioecological observations and up-to-date contamination mapping based on advanced geoinformation techniques give an ability to prepare for, respond to, and manage potential adverse effects from pollution with radionuclides and heavy metals. Hyperspectral satellite imagery provides potentially powerful tool for soil contamination detection and mapping. An intention to find a relation between remotely sensed hyperspectral and ground-based measured soil contamination fractions in area of the uranium mill tailings deposits near Kamianske city was made. An advanced algorithm based on known TCMI (target-constrained minimal interference)-matched filter with a nonnegative constraint was applied to determine the soil contamination fractions by hyperspectral imagery. The time series maps of spatial distribution of the soil contamination fractions within study area around the Sukhachevske tailings dump are presented. Time series analysis of the map resulted in two independent parameters: the average value for the entire observation period and the daily mean increment of the soil contamination fractions.

Keywords: Pre-Dnieper chemical plant, uranium mill tailings, soil contamination fractions, hyperspectral imaging, spectral unmixing, time series analysis

1. Introduction

Nowadays, agriculture became the leading branch of the Ukrainian economy. Ukraine places among the world's top 10 producers of wheat, barley, corn, and soybeans and is the leader in the production and export of sunflower. One-third of Ukrainian agrarian products are exported to European countries. Such agricultural achievements are primarily caused by the rich land potential of the country. There are 60.4 million hectares of mollisols in Ukraine. Moreover, almost 42 million hectares (about 69%) of them are arable lands and farmlands. The southern and central regions of Ukraine are especially fertile due to the highest content of humus in the local mollisols [1].

According to the Constitution of Ukraine, "Land shall be the main national asset and as such shall be under special protection" (Art. 14). Nevertheless, more than half of the fertile lands in Ukraine suffer from different types of degradation: loss of the fertile layer, erosion, and contamination. The problem of radioactive contamination of soils is extremely topical for Ukraine [2].

The mining and processing of uranium ore started in Soviet Ukraine in the late 1940s as a secret without any requirements of environmental safety. Uranium raw materials enrichment lasted in Ukraine until the Soviet Union fell apart in 1991. Currently, there are about 120 local contaminated sites that are identified inside and near Kamianske (former city name is Dniprodzerzhynsk). The exposure dose of these spots in several times is higher than the control.

Production Association of Pre-Dnieper chemical plant (PA PCP) was one of the most powerful uranium enterprises of the former USSR. Activity of the enterprise has led to formation of uranium ore traces. In particular, they can appear in a form of specific contamination fractions on the surface of the soil. Such focal points of the radiation pollution are enriched with radium-226 and other uranium-thorium radionuclides. They have been formed as a result of uranium extraction and ore recycling. There are nine radioactive waste storage facilities developed since the production at the PA PCP was launched. They are placed at the plant site as well as adjacent areas.

Radioactive solid and liquid waste is accumulated in gullies and hollows and on the upper terrace of the Dnieper river near other mining and metallurgical enterprises. Agricultural lands adjoin uranium tailings from all sides. Each year, as a result of wind erosion, more than 30 tons of radioactive dust settles on the arable lands. Also, radionuclides may be discharged to the atmosphere through by radon emanation.

Production of uranium concentrate was halted on PCP in connection with Soviet Union collapse. Several liquidation-reclamation actions were made within environmental programs to manage safe condition of uranium production with PCP [3, 4]. However, the management and maintenance of these neglected facilities are still problems of the current interest. Therefore, environmental security requires the use of modern technologies [5].

A set of environment security technologies are needed to be implemented to monitor, assess, and visualize the agricultural soil contamination. It will give an ability to prepare for, respond to, and manage potential adverse effects from radioactive pollution.

The goal of the research was the elimination of the negative environmental effects of the PCP activity. Uranium milling activities is a source of considerable soil contamination by radioactive substance, making a harmful impact on environment and the population. Regular radioecological observations and up-to-date contamination mapping become a primary issue for environmental protection of areas surrounding uranium ore milling works. Hyperspectral satellite imagery provides a potentially powerful tool for radioactive pollutant detection and mapping.

2. Main sources and ways of soil contamination

2.1. Uranium mill tailings deposits

The Sukhachevske tailings dump and Base S tailings deposit are located on the right bank of the Dnieper 5 km southeast of Kamianske (**Figure 1a**).

The Sukhachevske wet uranium tailings dump is still partly covered with water. The dump is located on the left branch of the Rozsoluvata ravine in the Sukha Sura river valley. The tailings' construction is a ravine partitioned by two dams and consisted of two sections, one arranged behind the other (**Figure 1b**). Base S tailings deposit is located 830 m east of the Sukhachevske tailings dump.

The chemical composition of wastes formed the major source of uranium ore minerals (quartz, feldspar, hydromica, kaolinite) combined with the chemicals used during ore processing (sulfuric and nitric acid) and products of neutralization of the resulting acidic environment (lime). The exposure dose rate of gamma radiation on the surface of the waste not covered by water varies from 100 to 1800 mcR/h. Power of equivalent dose of gamma radiation at a height of 1 m above the surface of the stored materials is changed in the downstream part from 0.31 to 4.08 μSv/h, in the upper reaches of 0.59–4.4 μSv/h, and averages the lower section of 1.91 ± 0.55 μSv/h.

The concentrations of radon-222 in the waste varies from 0.17 ± 0.03 MBq/m^3 to 1.73 ± 0.25 MBq/m^3 at a depth of 0.6 m. The flux density of radon-222 from the surface is in the range of 0.03–1.475 Bq/m^2·s.

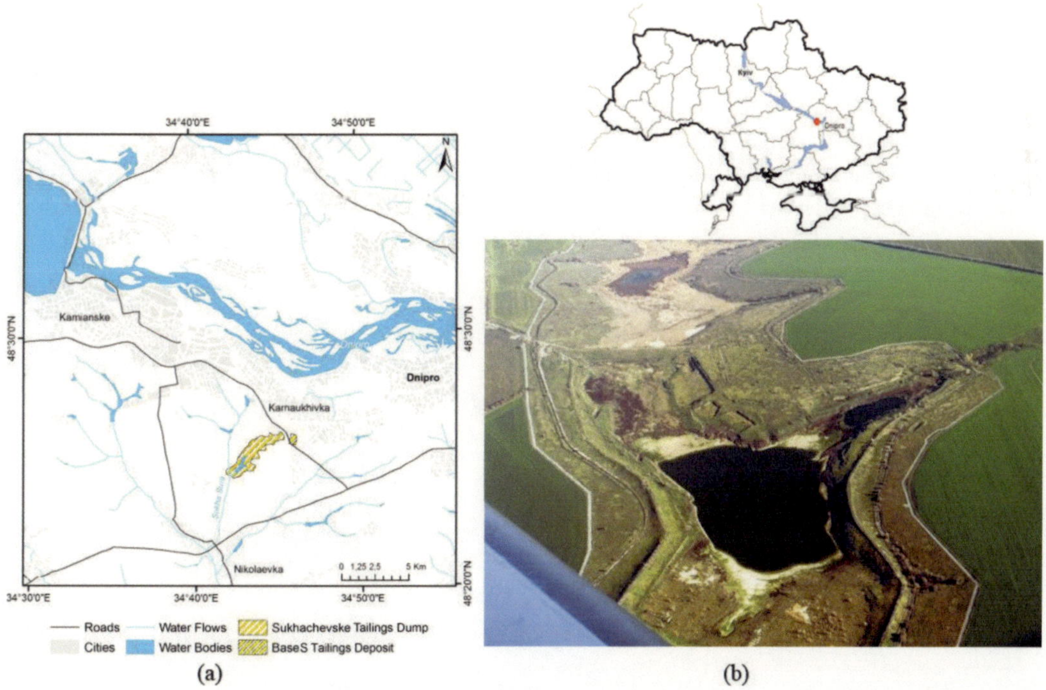

Figure 1. Location of the main uranium mill tailings deposits around Kamianske city (a) and an aerial photo of Sukhachevske tailings dump from http://wikimapia.org (b).

Liquid waste is represented by the waters of a pond in the central part of the tailings dump and pore water in the bulk of the solid waste. The slurry water contains barium, cadmium, strontium, cobalt, fluoride, nickel, zinc, silicium, manganese, and aluminum.

The main purpose of another tailings deposit—Base S—was a temporary warehousing of raw uranium delivered to the plant for further processing. It was built in 1960 and was in operation until 1990.

Now, the technology for waste storage at the tailings dump is disturbed: water pumping stations and conduits were destroyed; the feeding of water to the tailings dump was terminated; fencing facilities are destroyed in many places; and sewage water pipeline from PCP to the Sukhachevske uranium tailings dump suffered from numerous violations and corrosion. As a result, radioactive dust and sewage water have contaminated surrounding areas.

2.2. Ways of radioactive contamination spreading

2.2.1. Wind-driven spread of contaminants

The frequency of repetition directions of the wind is one of the determining factors to assess the transport of fine particles from the surface of waste storage. The right side of Kamianske city is under prevailing southerly wind spreading. This fact coincides with increment of technogenic load from the southern side of the second section of Sukhachevske tailings dump [6]. The predominant annual average wind frequency of occurrence has its maximum in north-northwest direction (**Figure 2**). However, the excess over the rest is minimal.

Average wind speed is in the range of 2.9–4.4 m/s (**Table 1**).

A large part of the surveillance zone near the Sukhachevske tailings dump is used for agricultural production. About 90% of the total area of farmland is plowed and used for cereals (wheat, barley, corn, oats), forages, and technical crops (sunflower and rape) cultivation. Lower slopes and bottoms of the ravines are used for haymaking and grazing. The forest vegetation in the area is represented by forest belts on the plateau and its slopes. Each year, because of wind erosion, considerable amount of contaminated dust particles settle on arable lands.

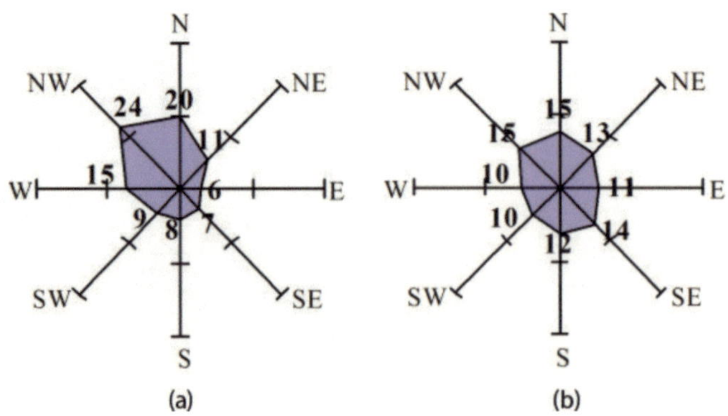

Figure 2. Frequency of wind directions (%) by long-term observations: (a) July average and (b) annual average [7].

http://dx.doi.org/10.5772/intechopen.72601

Month	I	II	III	IV	V	VI	VII	VIII	IX	X	XI	XII	Year
Wind speed (m/sec)	4.2	4.4	4.2	3.9	3.3	3.2	3.0	2.9	3.0	3.5	3.6	3.8	3.6

Table 1. Average month and annual wind speed by long-term observations (m/s) [7].

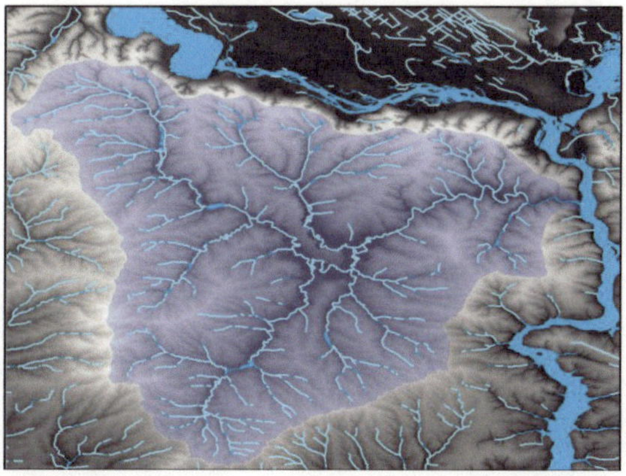

Figure 3. Basin of the Mokra Sura river.

2.2.2. Transfer of contaminants by hydrological network and surface runoff

Surface runoff from contaminated land is one of the major processes responsible for the contamination of water bodies. From the catchment area, the contaminants draining into the streams accumulate in floodplains and riverbeds. Rivers, drying out in warm seasons, cause secondary wind-driven contamination.

The Sukha Sura river is a tributary of the Mokra Sura river. Meantime, the pool of the Mokra Sura river is bordered with pool of the Samotkan river from the northwest, with the river Bazavluk in the southwest, with the Tomakovka river from the east, and near the Dnieper river from the north and east (**Figure 3**).

3. Materials and methods

3.1. Target and background spectral separation

Known algorithms for polluting agents' detection, which are used to analyze the spectrometric measurements, are based on the target and background spectral separation [8]. Therefore, the target spectra are necessary before starting the analysis of hyperspectral imagery. Before hyperspectral imaging engagement, the spectrometric measurements of contaminated soil samples were performed in ground control points (GCP), which are plotted in **Figure 4a**. Soil equivalent dose rate measurements were performed 10 times for each sample and then averaging (**Tables 2** and **3**).

Laboratory spectrometric measurements of field samples were carried out using the FieldSpec 3FR high-precision portable spectrometer (**Figure 4b**) developed by Analytical

Spectral Devices, Inc. (ASD, Inc.) company (http://www.asdi.com//). The main specifications of FieldSpec 3FR instrument are as follows: spectral range is 350–2500 nm, spectral sampling interval is 1.4 nm inside 350–1000 nm range, 2 nm inside 1000–2500 nm range, and output spectral sampling interval is 1 nm (interpolated).

Spectra of the field samples measured using FieldSpec3FR were calibrated as absolute reflectance, including both reflectance value and standard deviation. It is clear that to match the

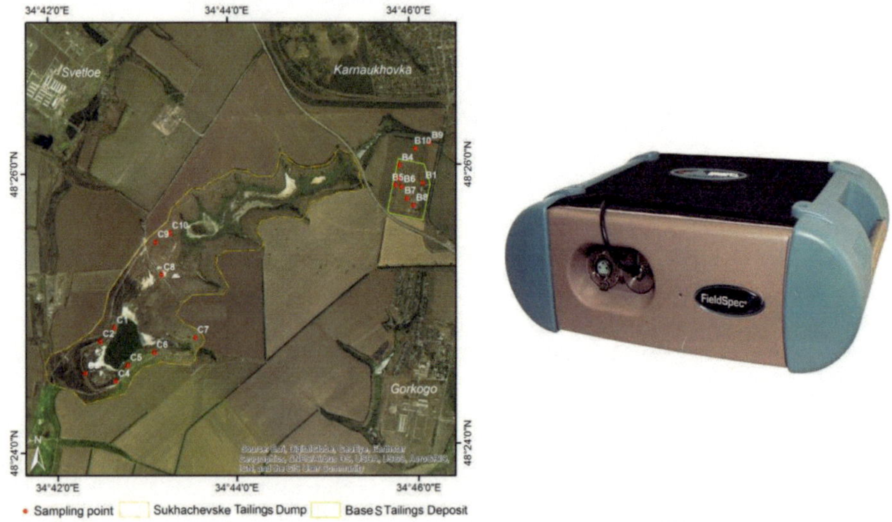

(a) (b)

Figure 4. Ground control points of in situ measurements (a) and ASD FieldSpec 3FR portable spectrometer used for spectrometric measurements of contaminated soil samples (b).

	The Sukhachevske tailings dump									
Point code	C 1	C 2	C 3	C 4	C 5	C 6	C 7	C 8	C 9	C 10
Equivalent dose rate (μSv/h)	0.11–0.16	0.10–0.14	0.09–0.12	0.12–0.16	0.11–0.14	0.08–0.12	0.09–0.13	0.69–0.80	0.16–0.19	0.18–0.23

Table 2. Results of in situ measurements of soil equivalent dose rate at the Sukhachevske tailings dump.

	The Base S tailings deposit									
Point code	B 1	B 2	B 3	B 4	B 5	B 6	B 7	B 8	B 9	B 10
Equivalent dose rate (μSv/h)	1.15–1.24	0.38–0.44	0.21–0.26	0.15–0.21	0.39–0.46	0.39–1.08	2.22–2.41	2.02–2.13	0.14–0.18	0.13–0.17

Table 3. Results of in situ measurements of soil equivalent dose rate at the Base S tailings deposit. Comparative to average background within Dniprodzerzhinsk city [7] equivalent dose rate in points B1, B6, B7 and B8 was 8-17 times more.

FieldSpec 3FR and Hyperion spectra correctly, the latter one should be recalibrated to land surface reflectance too.

The target spectra of soil samples at sites B and C within PCP acquired by FieldSpec 3FR spectrometer are shown in **Figure 5**.

3.2. Hyperspectral imagery time series and preprocessing

The Hyperion imaging spectrometer was part of the NASA's Earth Observing Mission 1 (EO-1). Hyperion was the first imaging spectrometer to routinely acquire science-grade data from Earth orbit. Its 242 bands covered the visible, near-infrared, and shortwave infrared bands (400–2500 nm) with 10 nm bandwidths. The spatial resolution of 30 m was sufficient to address most land application issues [9].

Images are available for free download through the United States Geological Survey (USGS) EarthExplorer. Timeline includes EO-1/Hyperion products on the 4-year period from 20 July 2012 to 13 July 2016 (**Figure 6**). All the images have been radiometrically calibrated based on *gain* and *offset* values from metadata file. As a result, a stack of 196 bands in 426–2395 nm range for each image have been formed.

The images were converted into surface reflectance using MODTRAN 4 atmospheric model approach [10]. It uses only approximate specification of sensor band locations (i.e., central wavelengths) and their radiometric calibration; no additional metadata were required [11].

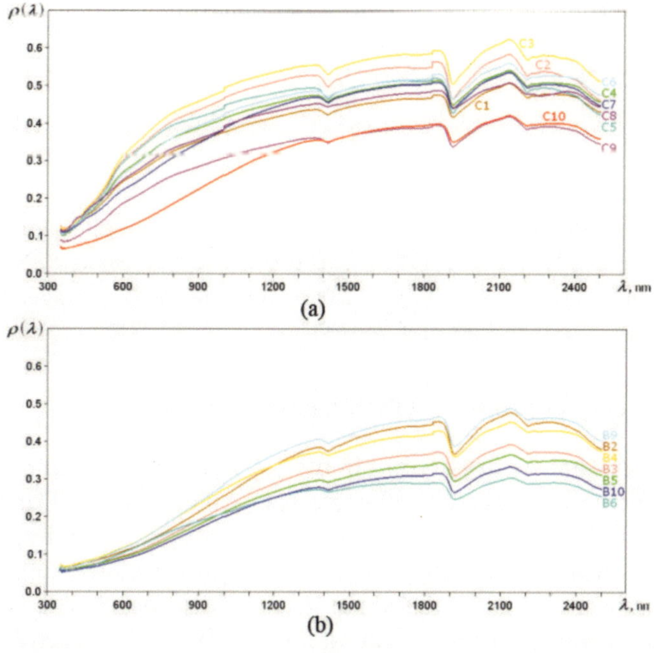

Figure 5. Target spectra of soil samples: (a) at the Sukhachevske tailings dump and (b) at the Base S tailings deposit.

Figure 6. EO-1/Hyperion images over the Sukhachevske tailings dump: Band combinations are *R*, 640; *G*, 548; and *B*, 487 nm.

Each band of every image of timeline was checked on salt-and-pepper noise in two iterations by signal-to-noise ratio (SNR). Bad band list was formed to remove low SNR bands [12]. New stack for images consists of 87 bands in 487–2324 nm range. All the images have been clipped by image coordinate system and geo-referenced.

3.3. Hyperspectral imagery processing for target spectral mapping

Usually, polluting substances are present in the soil in a small amount and therefore contribute insignificantly to the overall reflection spectrum. A spectral unmixing technique is used to detect such small impurities. Coarse spatial resolution of hyperspectral imager leads to the possibility of several different spectra capturing jointly, causing errors in their separation [13]. If the spectral samples are quite a few, it is possible to separate the different spectral one from another, even within a joint field of view. The linear unmixing model is used most commonly. This model calculates the contribution of each spectrum available within the field of view. Linear unmixing model provides the homogeneous spectral weighing in proportion to their fractions within the field of view [14]. Methods and algorithms for spectral unmixing are developed for decades [15]. In our case the hyperspectral imagery from NASA's EO-1/Hyperion satellite sensor (https://eo1.gsfc.nasa.gov/) was used. This instrument operates in 400–2500 nm spectral range with 30 m spatial resolution on the ground surface [16].

The general processing dataflow is described by the flowchart in **Figure 7** diagram.

Both hyperion sensor and FieldSpec spectrometer raw data are calibrated for land surface reflectance output. Then, the high spectral resolution FieldSpec data transformed into Hyperion's spectral signatures through the FieldSpec's spectral convolution with Hyperion's band spectral responses [17]. Now, it is possible to perform pixel-by-pixel matching of the target and current signatures for spectral fraction mapping. The matching procedure implies simultaneous reliability evaluation of the target signature retrieving. Further, these estimates are used to adjust the values of detected target spectral fractions (the fusion operator). Finally, the distribution maps of the target spectra are generated for the entire hyperspectral image scene.

The spectral unmixing model assumes that every *i*th spectral signature, $i = 1 \ldots n$, can be written as an *m*-dimensional vector y_i, where *m* is the number of spectral samples, and each *j*th target spectrum—by an *m*-dimensional vector x_j, $j = 1 \ldots p$. Let X as the matrix of target spectra of size $m \times p$, and $\alpha_i = (\alpha_1, \alpha_2 \ldots \alpha_p)^T$ as the vector of target spectral fractions in the *i*th spectral signature. Linear mixing model for y_i spectrum is expressed by equation:

$$y_i = X\alpha_i + \varepsilon_i \tag{1}$$

where ε_i is the residual vector which can be considered as the additive noise [18].

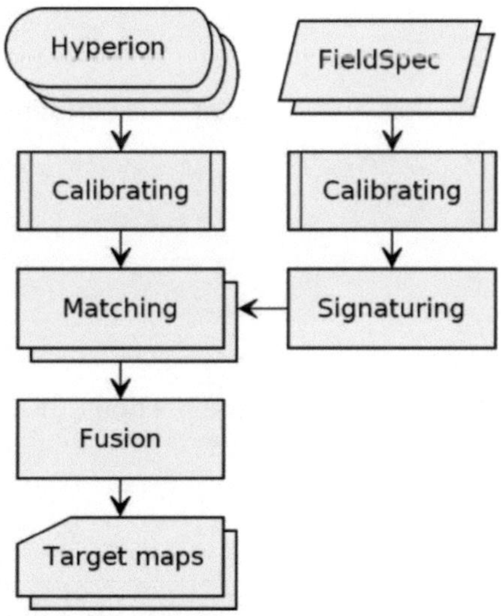

Figure 7. Hyperspectral imagery general flowchart.

The main restriction of unmixing is exceedance of the spectral sample number over the number of target spectra which are unmixed:

$$m \geq p \tag{2}$$

If all elements of X matrix are known, then the problem comes down to solving a system of linear equations by least squares (LS) method, possibly with some constraints: nonnegatively (nonnegatively constrained least squares (NCLS)), sum-to-one equality (sum-to-one constrained least squares (SCLS)), or both simultaneously (fully constrained least squares (FCLS)). In [8] paper a special algorithm based on modified FCLS method was proposed.

Unfortunately, in practice, as a rule, the all spectral composition of the whole scene is unknown. In this case the other method must be applied which extract one or more known target spectra, and the rest are considered as unwanted [19]. The TCMI (target-constrained minimal interference) filter, proposed in [20] paper, is the most perfect of such methods. In TCMI filter the estimation of sum of target spectral fractions in ith signature is equal to $w^T y_i$, where y_i is the mixed spectrum and w is the solution of minimization problem:

$$\begin{cases} w^T x_j = \begin{cases} 1 & \text{if j is target} \\ 0 & \text{otherwise} \end{cases} \\ j = 1 \,..\, p \\ \sum_{i=1}^{n} (w^T y_i)^2 \rightarrow \min \end{cases} \tag{3}$$

Generally, it is can be assumed that the first k spectra are the target and rest $k + 1 \ldots p$ ones are unwanted. Then, an explicit formula for w will be as follows:

$$w^T = (\underset{1}{1}..\underset{k}{1}, \underset{k+1}{0}..\underset{p}{0})\,(X^T Y^{-1} X)^{-1} X^T Y^{-1} \tag{4}$$

where $Y = \sum_{i=1}^{n} y_i y_i^T$.

If the target signature is alone, then the TCMI filter is simplified to the CEM (constrained energy minimization) one [21]. Thereupon,

$$w_{CEM}^T = \frac{x_1^T Y^{-1}}{x_1^T Y^{-1} x_1}$$

where x_1 is the target spectrum.

In order to apply the TCMI filter for target spectral fractions, the ones estimations in ith mixed signature will be.

$$\alpha_{TCMIi} = (X^T Y^{-1} X)^{-1} X^T Y^{-1} y_i \tag{5}$$

where α_{TCMIi} is a p-dimensional vector. The TCMI and CEM filters can be reduced to linear transform of spectra with the kind of a LS method also known as the OSP (orthogonal subspace projection) method [22].

Any components of α_{TCMIi} vector can be negative. Theoretically, the small value of fraction estimate indicates that this spectrum is not present inside mixed signature. Nevertheless, the negative values are unacceptable as fraction estimates. To avoid the negative values of estimates, it is reasonable to apply a method similar to the NCLS. Such improvement of the classical TCMI algorithm was proposed in [23] paper.

Combination of the TCMI and NCLS methods consists in finding a target spectral fractions α_i in the ith mixed signature as a minimum value in the equation system with constraints:

$$\begin{cases} (y_i - X\alpha_i)^T Y^{-1} (y_i - X\alpha_i) \rightarrow \min \\ \alpha_{ij} \geq 0,\ j = 1, 2 .. p \end{cases}$$

Like TCMI and CEM, the TCMI-NCLS method is reduced to spectral multiplication by the $Y^{-1/2}$ matrix and to succeed application of the NCLS method. The TCMI-NCLS algorithm provides a higher accuracy than the classical TCMI in spectral unmixing for land cover classification [23].

The processing of hyperspectral imagery for soil contamination mapping of study area was carried out through the determination of target spectral fractions in each hyperspectral pixel by the TCMI-NCLS algorithm.

Our previous experience has shown that it is necessary to estimate the error probability and to adjust the expected values of target spectral fractions (the Fusion procedure) for reliable results. Similar approach was applied for soil contamination mapping within Kiev city area using the hyperion hyperspectral image [24]. Techniques based on information divergence [25], Bhattacharyya statistical distance [26], spectral-topological classifier [27], and fuzzy decision tree [28] were considered. Finally, the information divergence and Bhattacharyya statistical distance were involved to adjust the target spectral fractions after TCMI-NCLS algorithm applying over input hyperspectral image.

It is possible to ensure further improvement in evaluation reliability by analyzing the time series of research area imagery rather than stand-alone images. Time series analysis is a universal tool for the systems and process state assessment, as well as for its prediction. Time series analysis is especially important for the remote sensing data processing [29]. The purpose of time series analysis is to determine the parameters of the occurring change dynamics, primarily the trend and periodic components [30]. The Earth's surface imagery time series should be considered as composite of individual time series in each pixel.

4. Results and discussion

In the current research, the subfractions of each target spectrum were summed up, and in this way, the maps of spatial distribution of soil with pollution has been formed. **Figure 8** contains the spatial distribution of the total fractions of target spectra within the scene of the study area. Since the applied algorithm detects contaminant fractions only on the surfaces of exposed soil and has restrictions for detecting them on other soils covered with snow or vegetation, spatial distribution of the total fractions of target spectra is differently represented in the images of certain stages of the season. The least indicative are the images of the warm period when most of the study area is covered by natural (in the floodplain of the Sukha Sura river) or agricultural (on the arable lands) vegetation as it can be seen from the images of 10 May 2013, 20 July 2013, and 13 July 2016. The image for the snowless winter period (18 February 2015) demonstrates spatial distribution of the contaminant fractions most entirely, excepting small areas of artificial surfaces and areas covered with snow and ice. Spatial distributions of the contaminant fractions on the images of the autumn period (18 September 2014, 17 September 2015) are reflected depending on the soil exposure and the way it was cultivated during agricultural works and differ from year to year. Besides, the detection of contaminant fractions can also be affected by the state of the soil at the time of survey, to the extent that it is waterlogged, cultivated, or eroded.

The pixel-by-pixel simultaneous processing of all target spectral maps resulted in spatial distributions of time series parameters. Because the total number of hyperspectral images in time series was not too large, an analysis of their time series turns into the linear trend extracting [31]. The linear trend is described by two independent parameters: the average value for the entire observation period and the daily mean increment. The results of parameter calculation of Hyperion image's time series linear trend for both scenes are illustrated in **Figure 9**.

The data on average value and daily increment show hot spots of high technogenic load around Rozsoluvata ravine.

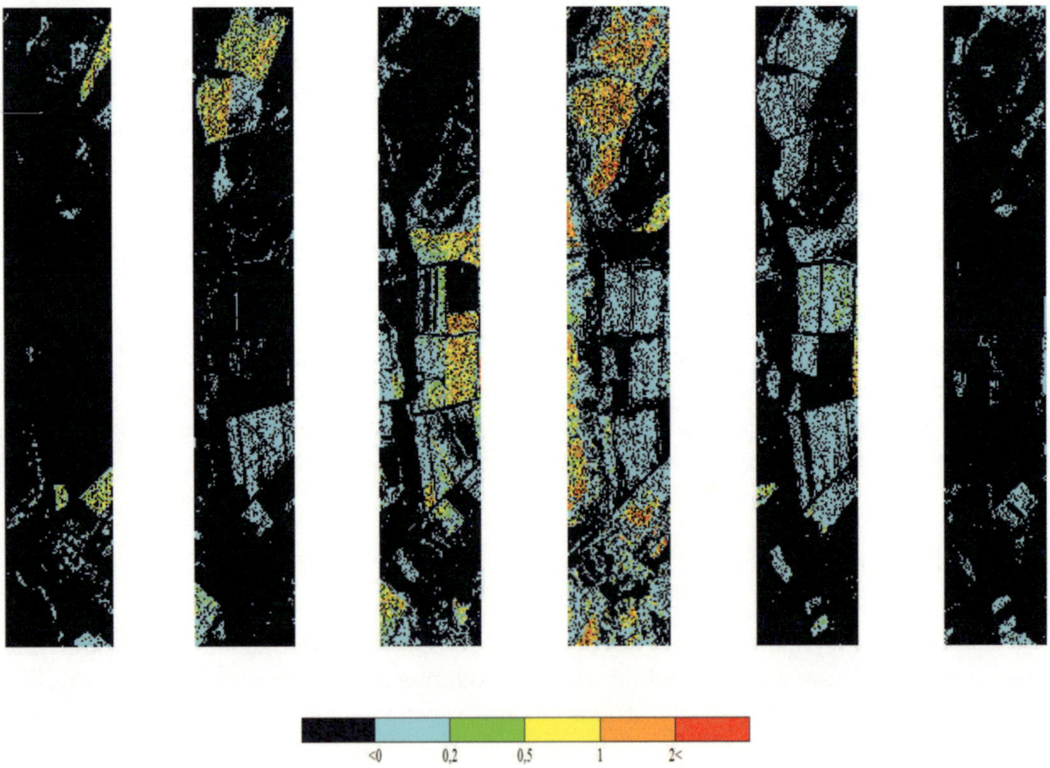

Figure 8. Target spectral total fractions by fused TCMI-NCLS algorithm.

http://dx.doi.org/10.5772/intechopen.72601

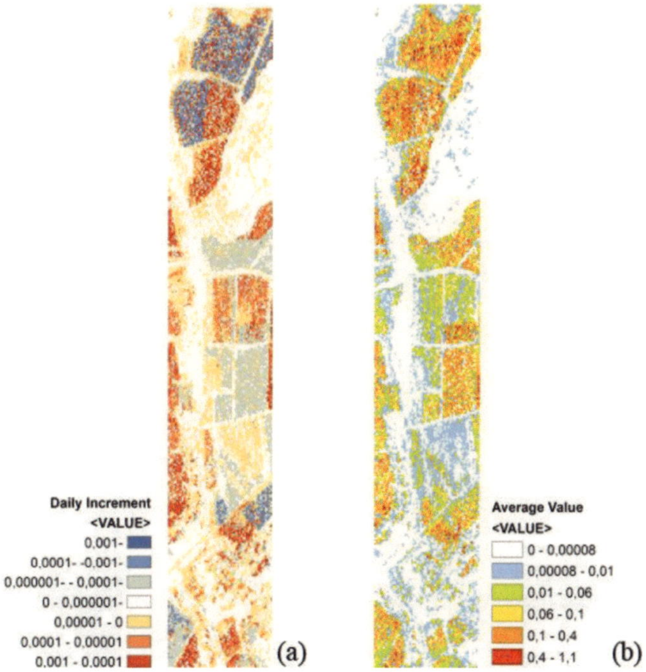

Figure 9. Spatial distributions of the time series parameters: (a) daily mean increment and (b) average value for the entire observation period.

5. Conclusions

Our research was made with intention to find a relation between remotely sensed hyperspectral and ground-based measured soil contamination fractions in the area of the uranium mill tailings deposits. Other types of urban industrial landscapes were not involved in the process of hyperspectral classification. Airborne dust and erosion processes were selected as the main reasons of environment pollution with radionuclides and heavy metals in the territory occupied with two tailings of uranium mill tailings. The spectra of field samples taken near the two deposits were compared with hyperspectral images. The maps on average value and daily increment assessment are background to classify area with different levels of technogenic load.

Additionally, our research has confirmed that hyperspectral imaging is a useful and an efficient tool for soil contamination mapping. One allows to detect small contaminant fractions on the soil surface by spectral end-member unmixing, if it is not shaded by vegetation or other covers. The proposed NCLS-TCMI algorithm is more advanced than the known CEM and TCMI ones, and it provides more reliable detection of soil contaminant's fractions. At the same time, the similarity of contaminated and non-contaminated soils' spectra and the small value of detected fractions have resulted in the need for additional adjustment of mapping outputs. This specified problem can be mitigated by taking certain measures when preparing the input data and carrying out the mapping. First, the reference spectra should be prepared not only for soils but also for other land covers within the scene. This will make it possible to carry out a full-scale land cover classification and to build a mask of soil of interest only before the final mapping. Second, the reference spectra of both contaminated and non-contaminated soils are required for contaminants' reliable detection. This will permit the Bayesian rule engagement for similar spectral discrimination.

Future works should be devoted to the development of complete all-in-one technology for mapping of soil contamination using hyperspectral imagery and its wide-ranging statistically significant probation over variety of test sites.

Author details

Sergey A. Stankevich[1*], Mykola M. Kharytonov[2], Anna A. Kozlova[1], Vadym Yu. Korovin[3], Mykhailo O. Svidenyuk[1] and Alexander M. Valyaev[3]

*Address all correspondence to: st@casre.kiev.ua

1 Scientific Centre for Aerospace Research of the Earth, NAS of Ukraine, Kiev, Ukraine

2 Dnepropetrovsk State Agrarian and Economy University, Dnipro, Ukraine

3 Institute of Geotechnical Mechanics, NAS of Ukraine, Dnipro, Ukraine

References

[1] Kharytonov MM, Pashova VT, Bagorka MO, Kozechko VI, Dudar TO. Arable lands degradation in the northern steppe zone of Ukraine. Agriculture and Forestry. 2016;**62**(2):71-80. DOI: 10.17707/AgricultForest.62.2.05

[2] Kharytonov M, Bagorka M, Gibson P. Erosion effects in the central steppe chernozem soils of Ukraine. I. Soil properties. Agricultura [Internet]. 2004;**3**(1):12-18. ISSN 1581-5439. Available from: http://www.agriculturaonline.com/portal [Accessed: October 18, 2017]

[3] Voigt G, Fesenko S, editors. Remediation of contaminated environments. In: Baxter MS, editor. Radioactivity in the Environment. 1st ed. Amsterdam: Elsevier Science; 2009. 14. 496 p. ISBN: 978-0-08-044862-6

[4] Smičiklas I, Šljivić-Ivanović M. Radioactive contamination of the soil: Assessments of pollutants mobility with implication to remediation strategies. In: Marcelo LL, Sonia S, editors. Soil Contamination – Current Consequences and Further Solutions. Rijeka: InTech [Internet]; 2016. pp. 253-276. DOI:10.5772/64735. Available from: https://www.intechopen.com/books[Accessed: October 18, 2017]

[5] Stankevich SA, Kharytonov NN, Dudar TV, Kozlova AA. Risk assessment of land degradation using satellite imagery and geospatial modelling in Ukraine. In: Kaswamila A, editor. Land Degradation and Desertification – A Global Crisis. Rijeka: InTech; 2016. pp. 53-77. DOI: 10.5772/61629

[6] Kharytonov M, Benselhoub L, Shupranova A, Kryvakovska R, Khlopova V. Environmental assessment of atmospheric pollution in Dnipropetrovsk province (Ukraine). Studia Universitatis "Vasile Goldis". 2015;**25**(2):125-130. ISSN: 1584-2363

[7] Environmental passport of Dneprodzerzhinsk city. Dnepropetrovsk; 2003. p. 144

[8] Stankevich S. Land-cover classification on hyperspectral aerospace images by spectral end members unmixing. Journal of Automation and Information Sciences. 2006;**38**(12):31-41. DOI: 10.1615/J Automat Inf Scien.v38.i12.40

[9] Pearlman J, Barry PS, Segal CC, Shepanski J, Beiso D, Carman SL. Hyperion, a space borne imaging spectrometer. IEEE Transactions on Geoscience and Remote Sensing. 2003;**41**(6):1160-1173. DOI: 10.1109/TGRS.2003.815018

[10] San BT, Suzen ML. Evaluation of different atmospheric correction algorithms for EO-1 Hyperion imagery. International Archives of the Photogrammetry, Remote Sensing and Spatial Information Science. 2010; **XXXVIII**(8):392-397. EID: 2-s2.0-84877874959. Available from: http://www.isprs.org/proceedings/XXXVIII/part8/pdf/W03C05_20100306205904.pdf [Accessed: October 28, 2017]

[11] Bernstein L, Jin X, Gregor B, Adler-Golden S. Quick atmospheric correction code: Algorithm description and recent upgrades. SPIE Optical Engineering. 2012;**51**(11):111719-1-111719-11. DOI: 10.1117/1.OE.51.11.111719

[12] Lukin V, Ponomarenko N, Fevralev D, Vozel B, Chehdi K, Kurekin A. Classification of pre-filtered multichannel remote sensing images. In: Escalante-Ramirez B, editor. Remote Sensing – Advanced Techniques and Platforms. Rijeka: InTech Open; 2012. p. 75-98. DOI:10.5772/1808. Available from: https://www.intechopen.com/books[Accessed: October 28, 2017]

[13] Melesse AM, Weng Q, Thenkabail PS, Senay GB. Remote sensing sensors and applications in environmental resources mapping and modeling. Sensors. 2007;**7**(12):3209-3241. DOI: 10.3390/s7123209

[14] Nielsen AA. Spectral mixture analysis: Linear and semi-parametric full and iterated partial unmixing in multi- and hyperspectral image data. International Journal of Computer Vision. 2001;**42**(1-2):17-37. DOI: 0.1023/A:1011181216297

[15] Eismann MT, Hardie RC. Stochastic spectral unmixing with enhanced endmember class separation. Applied Optics. 2004;**43**(36):6596-6608. DOI: 10.1364/AO.43.006596

[16] Ungar SG, Pearlman JS, Mendenhall JA, Reuter D. Overview of the earth observing one (EO-1) mission. IEEE Transactions on Geoscience and Remote Sensing. 2003;**41**(6):1149-1159. DOI: 10.1109/TGRS.2003.815898

[17] Cundill SL, van der Werff HMA, van der Meijde M. Adjusting spectral indices for spectral response function differences of very high spatial resolution sensors simulated from field spectra. Sensors. 2015;**15**(3):6221-6240. DOI: 10.3390/s150306221

[18] Popov MA, Stankevich SA, Moldovan VD. Artificial objects detection by subpixel processing of hyperspectral aerospace imagery (Ukrainian). Journal of State Aviation Research Institute Kyiv; December. 2006;**9**(2):194-204

[19] Chang C-I. Hyperspectral Imaging: Techniques for Spectral Detection and Classification. N.Y: Kluwer Academic/Plenum Publishers; 2003. p. 396. DOI: 10.1007/978-1-4419-9170-6

[20] Kwan C, Ayhan B, Chen G, Wang J, Ji B, Chang C-I. A novel approach for spectral unmixing, classification, and concentration estimation of chemical and biological agents. IEEE Transaction on Geoscience and Remote Sensing. 2006;**44**(2):409-419. DOI: 10.1109/TGRS.2005.860985

[21] Farrand WH, Harsanyi JC. Mapping the distribution of mine tailings in the Coeur d'Alene River valley, Idaho, through the use of a constrained energy minimization technique. Remote Sensing of Environment. 1997;**59**(1):64-76. DOI: 10.1016/S0034-4257(96)00080-6

[22] Harsanyi JC, Chang C-I. Hyperspectral image classification and dimensionality reduction: An orthogonal subspace projection approach. IEEE Transactions on Geoscience and Remote Sensing. 1994;**32**(4):779-785. DOI: 10.1109/36.298007

[23] Stankevich SA, Shklyar SV. Advanced algorithm for endmembers unmixing on hyperspectral image (Ukrainian). In: Proceedings of the 1st Ukrainian Conference with International Participation; May 2008. Kyiv: Naukova Dumka; 2008. pp. 85-89

[24] Popov MA, Stankevich SA, Lischenko LP, Lukin VV, Ponomarenko NN. Processing of hyperspectral imagery for contamination detection in urban areas. In: Alpas H, Berkowicz SM, Ermakova IV, editors. Spinger. Environmental Security and Ecoterrorism. 2011. pp. 147-156. DOI: 10.1007/978-94-007-1235-5

[25] Cover TM, Thomas JA. Elements of Information Theory. New York: Wiley; 1991. 576 p. ISBN 0-471-20061-1

[26] Fukunaga K. Introduction to Statistical Pattern Recognition. San Diego: Academic Press; 1990. 616 p. ISBN 0-12-269851-7

[27] Stankevich SA. Algorithm for statistical classification of remote sensing objects by their spectral-topological features (Ukrainian). In: Proceedings of Scientific Bulletin of National Mining University. Dnepropetrovsk. 2006. 7. pp. 38-40. ISSN 2071-2227

[28] Stankevich SA, Levashenko VG, Zaitseva EN. Multispectral satellite imagery classification using fuzzy decision tree. Communications. Žilina, Slovakia: University of Žilina. 2014;**16**(1):109-113. ISSN: 1335-4205

[29] Hostert P, Griffiths P, van der Linden S, Pflugmacher D. Time series analyses in a new era of optical satellite data. In: Status Quo and the Pathway Ahead. Kuenzer C, Dech S, Wagner W, editors. Remote Sensing Time Series Revealing Land Surface Dynamics: Cham. Springer, Cham; 2015. 22. pp. 24-42. DOI: 10.1007/978-3-319-15967-6_1

[30] Yaffee R, McGee M. An Introduction to Time Series Analysis and Forecasting. 1st ed. San Diego: Academic Press; 2000. p. 528

[31] Verbesselt J, Hyndman R, Newnham G, Culvenor D. Detecting trend and seasonal changes in satellite image time series. Remote Sensing of Environment. 2010;**114**(8):106-115. DOI: 10.1016/j.rse.2009.08.014

Section 5

Hyperspectral Imaging Methods

Chapter 8

Using the Target-Visualization Method to Process Hyperspectral Images

Tatiana A. Sheremeteva, Gennadiy N. Filippov and
Andrey M. Malov

Additional information is available at the end of the chapter

http://dx.doi.org/10.5772/intechopen.72249

Abstract

Processing hyperspectral images allows you to decode images and recognize objects in the scene on the base of analysis of spectrums. In some problems, information about the spectra may not be sufficient. In this case, visualization of data sets may use, for object recognition, by use additional non-formalized external attributes (for example, indicating the relative position of objects). Target visualization is a visualization adapted to a specific task of application. The method discussed in this chapter uses a way to visualize a measure of similarity to the sample. As a result of the transformation, the hyperspectral (multichannel) image is converted into a single-channel synthesized image in grayscale, on which the objects of interest for the problem under consideration are selected. By changing the brightness and contrast of the synthesized image, it is possible to interactively adjust the results of automatic processing.

Keywords: hyperspectral image, visualization, interpretation, detection, interactive participation

1. Introduction

Hyperspectral image (HSI) is a set of images of one object (scene, slide, field of view, etc.), that match different narrow wavelength subranges, provides powerful tool for detection and recognition of objects under surveying.

Hyperspectral data use is a serious problem caused by need of processing of huge amount of information (tens and hundreds gigabytes). Most of hyperspectral image recognition algorithms are highly resource-intensive, especially when striving to provide full classification of hyperspectral information. Information about spectra may be insufficient for some tasks. In those cases acceptable interpretation cannot be realized without human assistance. The data visualization facilitates making decision for an expert-user allowing him to use unobvious signs obtained from experience. Watching all of the band images from HSI (HSI can consist of more than 300 spectral bands) is a labor-intensive and inefficient task. Segmentation methods are used to represent visual information in more compact view. These methods mark out homogeneous regions in the image that allows representing hyperspectral image in view of some synthesized image, for example, in pseudo-color, which facilitates visual perception. However, only in some cases a criterion of true segmentation exists. Visualization only allows us to reveal errors. And for the result correction, we need to process all over again using altered parameters of the algorithm. When solving a certain task, not all of the objects are of equal interest. Obviously, the visualization should be done so that the objects of interest were represented in the result image in the best way, but the details that prevent from good visual perception were eliminated. I.e. visualization should be targeted. Such adaptation of data processing to a specific practical task can be realized by means of target visualization method. As a result of such transformation the information contained in HSI is represented in the synthesized image in grayscale where the brightness range is concentrated on the details of the image that are of interest for user. The method combines mathematical computation speed with visual assessment and possibility to correct interpretation process interactively. In addition, the useful information is appreciably compressed, that is very important for remote sensing tasks.

2. Target visualization of HSI

Target visualization converts a set of images that constitute a hyperspectral image into a synthesized single-channel image that provides a visual representation of the data for a particular application. To perform the target visualization of HSI the image transform method is used. The method visualizes measure of similarity to a sample within a space of attributes [1, 2, 7].

Hyperspectral image, or so-called hypercube, is a collection of pictures of a single object that correspond to different narrow subranges of wavelengths,

$$B \in \{B_{kmn}, k = 1, \ldots, K\}, m = 1, \ldots, M, n = 1, \ldots, N \tag{1}$$

where k – the spectral subrange number; (m, n) – the spatial coordinates of an image pixel; M – the number of lines in the image; N – the number of columns; B_{kmn} – the brightness value at point (m, n) in the *k*th frequency subrange.

As the attributes, that distinguish the objects of interest, the spectral characteristics are used. Moreover, can be used other characteristics that are used in tasks on image classification [3, 4], such as: features of brightness, texture, gradient and characteristics of adjacency. The values of the characteristics used for each pixel, can be obtained a priori or via image pre-processing.

Thus, each pixel of the HSI with certain spatial coordinates has its coordinates within a feature space. In the visualization method for each task on HSI processing a decision function is formed within a set of feature values. The decision function quantitatively determines belonging of the pixel to the object of interest. Then the result image is synthesized. Pixel brightness of this image conforms to the decision function value. Changing brightness range in the obtained image and using his notion about the object being studied, user can interactively set threshold to mark out the object of interest.

The main problem of the method is to choose a decision function that corresponds to a specific applied task. Some of tasks and the choice of the decision function for them are listed as an example below:

1. To pick out objects on the HSI that have spectral characteristics that are similar to characteristics O_k from a spectral library.
2. To pick out objects on the HSI that are similar to the characteristics of an object-sample with specified spatial coordinates—i.e., sample O_k is specified by spatial coordinates (m, n): $O_k = B_{kmn}$.
3. To pick out objects which have the most widely represented spectra on the HSI (endmembers).
4. To pick out objects which have the most uncommon spectra.
5. To pick out boundaries of homogeneous regions on the HSI.
6. To pick out objects on thematic images—i.e., on images obtained by using mathematical operations with spectral bands [5].

The suitable decision function for the tasks 1–5 is the measure of similarity to the sample [1, 2]. Similarity can be specified by one of the measures of closeness of the vectors used for HSI classification, such as:

- the Euclidean distance

$$d1_{mn} = \sqrt{\sum_{K=1}^{K} \left(B_{kmn} - O_k\right)^2} \; ; \tag{2}$$

- the distance measured by the spectral angle

$$d2_{mn} = \arccos\left(\sum_{K=1}^{K} \left(B_{kmn} - O_k\right) / \sqrt{\sum_{K=1}^{K} B_{kmn}^2 \sum_{K=1}^{K} O_k^2}\right) \; ; \tag{3}$$

- the maximum distance

$$d3_{mn} = \max\left\{ |B_{1mn} - O_1|, \ldots, |B_{Kmn} - O_K| \right\}. \tag{4}$$

As a sample, multidimensional feature vector is chosen from set of points of the attribute space. A sample can be chosen either by specifying of a priori known feature values or by pointing to one of the discriminated objects.

Some set of vectors over region E of the feature space can be chosen as a sample, too. In this case it is possible to use the Mahalanobis distance—i.e., the distance given by the expression:

$$d4_{mn} = \left(B_{kmn} - \overline{O}\right)^{T} C^{-1} \left(B_{kmn} - \overline{O}\right), \tag{5}$$

where $\overline{O}$—the mean of set E, and C—a covariant matrix of set E.

For task 6, the suitable decision function is

$$d5_{mn} = f(B_{kmn}) - Por, \tag{6}$$

where $f(B_{kmn})$—some function of the spectral characteristics, used when synthesizing index images, and *Por*—the fixed a priori value of the index.

If an information on probability distribution of attributes of the sample is available, a measure based on conditional probabilities can be used (Bayesian approach). Any of a priori estimated characteristics of studied objects can be participant in forming of a measure.

By choosing of a sample and a measure of similarity we can adapt visual representation to a specified task on interpretation. Choice of a sample can be based on:

- visual analysis of some of the presented pictures;
- a priori information about the object of interest location;
- an information about the spectral features of the required object from a specially created database.

The value of the decision function (similarity measure value) is brightness of the pixel on the synthesized image. The largest brightness value is assigned to the pixels which are closest to the sample within the attribute space. The other pixels have brightness value according to their distance from the sample. By representing the matrix of the similarity measure in gradations of gray, we obtain a visualization of the HSI adapted to the applied task. Since the brightness of each image pixel characterizes its similarity to the object of interest, by visually choosing the brightness threshold and by varying the contrast, a specialist can discriminate the object that interests him by using his knowledge of the object and accumulated experience.

Various versions of the visualization of the same scene can be obtained by selecting various samples and similarity measures. This allows the observer to analyze the imaged scene by changing samples and similarity measures. Such a tool can be compared to a magnifying glass in feature space (on synthesized images various parts of the *feature* space can be detailed).

Target visualization used in various application areas of image processing, has a specificity that is determined by the area of the study and conditions of registration of images.

3. Target visualization of biomedical images

Computer image analysis is one of the relatively new medical technologies that are of great importance in medical research and diagnosing diseases. Due to the complexity of biomedical images, it is almost impossible to completely replace the human brain in processing them. At the same time, the method of target visualization of images allows, with the participation of a qualified user, to quickly identify and measure objects of interest.

An example of the visualization of hyperspectral images with different samples is shown in **Figures 1** and **2**. The source data was obtained using autofluorescent microscopy in the spectral range from 420 to 750 nm with an interval of 6 nm. (The image represents *Convallaria* rhizome section, acquisition with the Leica TCS SP5. The data were provided us by Biotechnologisches Zentrum der TU Dresden). **Figure 1** shows four of the 50 images of the specimen as an example of the initial data.

The capabilities of visualizing multispectral images are demonstrated in **Figure 2**. The results obtained by transforming the 50 initial images using various samples are shown in this figure. The samples are represented in the form of graphs of spectra under the transformed images.

Another example of using visualization of hyperspectral images of biomedical theme. This example demonstrates a non-invasive method of isolation of tumor cells using a hyperspectral image. In this paper we used the HSI represented on the website of the Australian National Agency CSIRO—www.cmis.csiro.au/iap.

In **Figure 3** the multispectral image of a mouse lung with a tumor, marked with fluorescent protein, obtained in 16 spectral bands is shown. It is assumed that the tumor has uncommon spectrums relating to the attribute space. The initial multispectral image is difficult for visual analysis (it is not easy to notice differences), but on the resulting synthesized images, shown in **Figure 4**, various regions of the studied object are clearly distinguished. For distinguishing these regions all of the initial spectral band images are involved into processing.

Under each synthesized image in **Figure 4**, there is the histogram which corresponds to the distribution of distances (d) in the attribute space from the spectrum of the sample to the spectrum of each pixel in the image. The histogram provides information on the number of pixels with a spectrum similar to the spectrum of the sample. Using the histogram, it is possible to set the start threshold value for the image binarization and distinguishing the objects like the sample.

Figure 5 shows the result of selecting objects corresponding to the three samples.

The thresholds were chosen at the first minimum of the histograms.

By altering the threshold value, it is possible to vary result of the visualization in according to the contextual information and experience of the user.

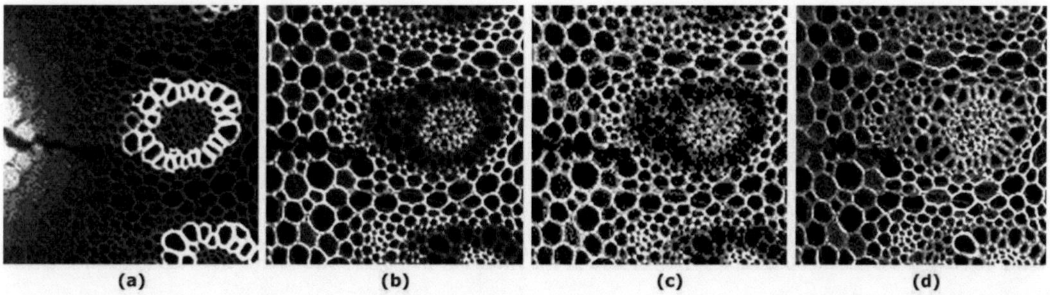

Figure 1. Sample 4 of 50 raw pictures from the HSI of micropreparation in the wavelength range 420–750 nm used for processing. (a) Wavelength of 462 nm, (b) 606 nm, (c) 654 nm, and (d) 702 nm.

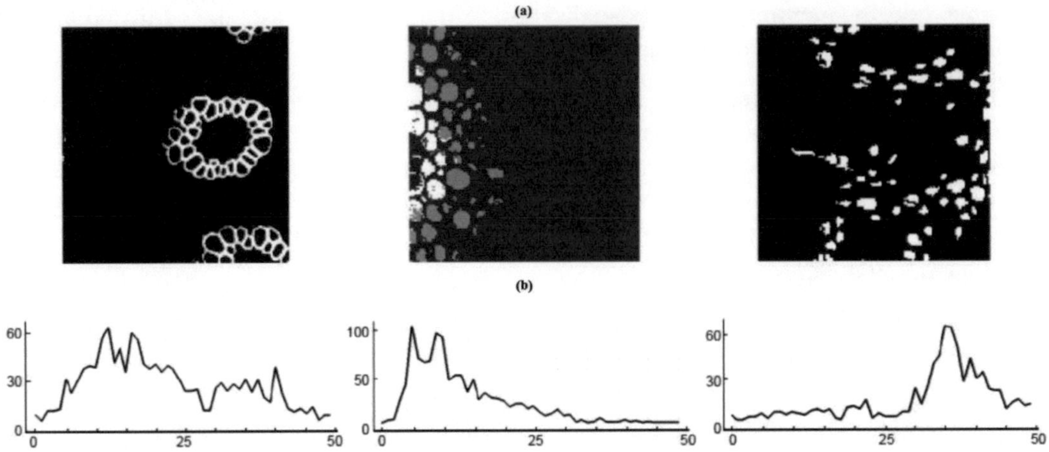

Figure 2. The image (a) was obtained by transforming the 50 source images using three appropriate samples (b).

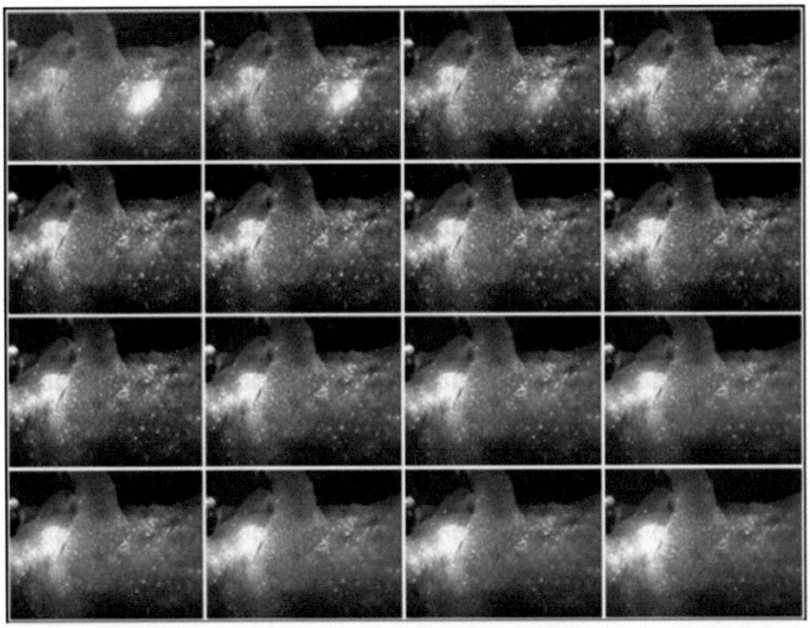

Figure 3. Multispectral image of a mouse lung obtained in 16 different spectral bands.

The method also allows us to distinguish pixels with spectral characteristics which are rare relatively most spectra of this image. For this purpose the Mahalanobis distance for some region E can be used—i.e., the distance from the vector **y** to the mean vector m_X of a set of vectors **X** which is specified by expression:

$$d(y, m_X) = (y - m_X)^T C_X^{-1} (y - m_X) \qquad (7)$$

where C_X is a covariant matrix of set **X**.

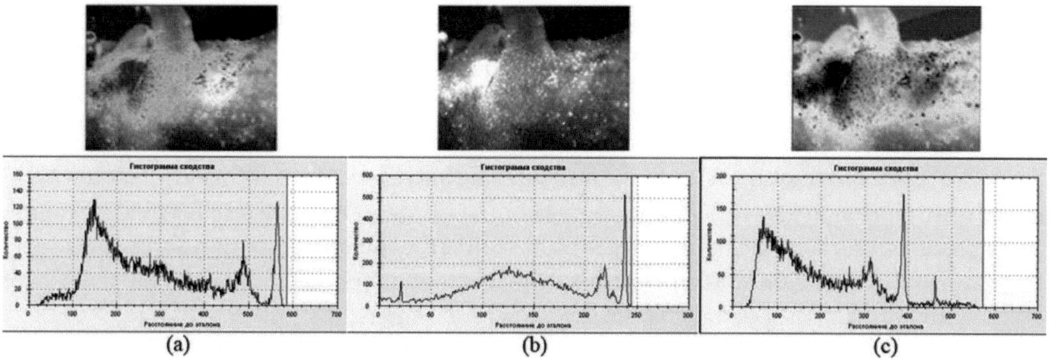

Figure 4. The result of visualization of 16 images for three samples: (a) Sample 1; (b) Sample 2; and (c) Sample 3.

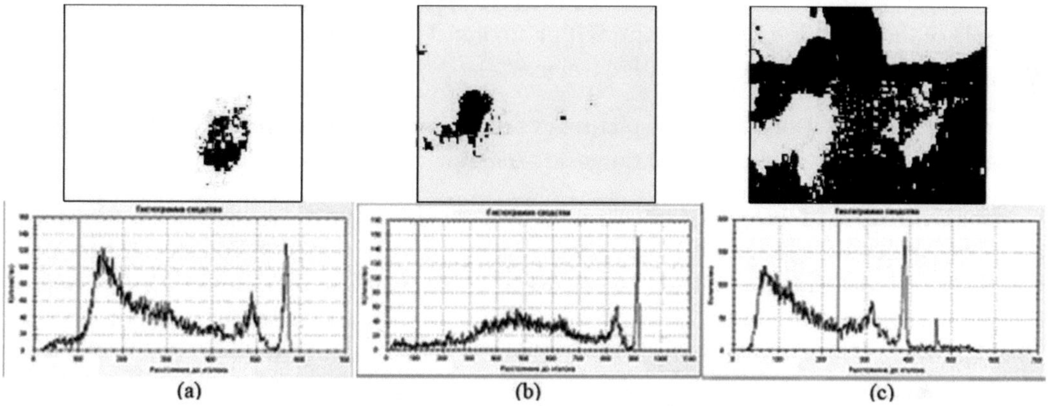

Figure 5. The result of selecting objects corresponding to the three samples (the images are inverted): (a) Sample 1; (b) Sample 2; and (c) Sample 3.

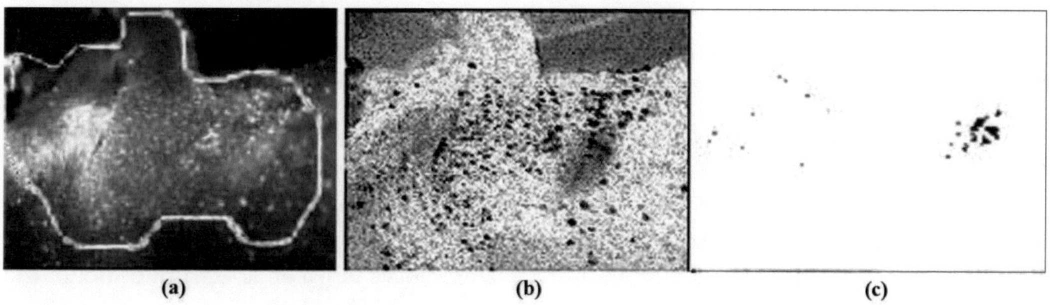

Figure 6. Steps of distinguishing pixels with uncommon spectral characteristics: (1) the spatial region defining the set **X**; (2) visualization of the initial images using Mahalanobis distance; (3) the result of distinguishing pixels with uncommon spectral characteristics (the image is inverted).

In our case, the set **X** is the set of vectors of the attribute space, which correspond to the most representative part of the image in **Figure 6a**. This area is indicated with the solid line. **Figure 6b** represents the result of visualization with the Mahalanobis distance over the set **X** use.

All pixels which are close by their spectral characteristics to m_X have light hue. Pixels, which are appreciably different from m_X, have dark hue. Nevertheless, the dark pixels may have different spectral characteristics. To find regions of the image, which are similar by uncommon characteristics, one should choose a dark pixel within the interesting region and use its spectral characteristics as a sample. In **Figure 6c**. the result of distinguishing one of the clusters of pixels with uncommon spectral characteristics is represented.

4. The target visualization of HSI using contextual information

The use of contextual information when processing hyperspectral images will be shown in the example of the detection of tampering with a financial document. The HSI of the document was presented to us by ZAO "Scientific Devices".

Figure 7 represents 4 of 125 initial pictures used for processing of the HSI of the financial document, obtained within spectral range 401–998 nm.

In these pictures, as in the others, not shown in **Figure 7**, it is difficult to separate the details of the text written with, though similar by spectrum, but still different ink. To detect the falsification the target visualization of the HSI consisting of 125 was used. To detect the falsification the target-visualization of the HSI consisting of 125 pictures was used. The goal of the visualization initially was to distinguish objects that have a spectrum similar to that of the main handwritten text.

The method [2] was used; the letter *o* (indicated in **Figure 8a** by an arrow) was chosen as a specimen of the main text. The choice of this sample was based on comprehension of the task (in this part of the text the falsification was senseless). Spectral angle measure $d2_{mn}$ was chosen as a measure of the similarity. This gave the synthesized image shown in **Figure 8b**. The brightness of each pixel of the synthesized image is determined by the similarity of the spectrum that corresponds to a specified pixel with the spectrum of the sample. The image shown in **Figure 8c** was obtained by varying the brightness and contrast.

A section of the text was then selected that did not appear on the synthesized image (indicated in **Figure 8a** by an arrow). Choice of this specimen was based on the fact that the spectrum of this object was different from the main handwritten text and was handwriting, too (another sort of inks might be used). The spectrum of this section, differed from the spectrum of the main text, was taken as a sample of an forgery. Differences between spectrum of ink of the main text and one's spectrum of the forgery are shown in **Figure 9**.

Visualization of the text that corresponds to this spectrum is shown in **Figure 10b** and after correcting the brightness and contrast—in **Figure 10c**. The letters "TY" and the numeral "0"belong to the forgery.

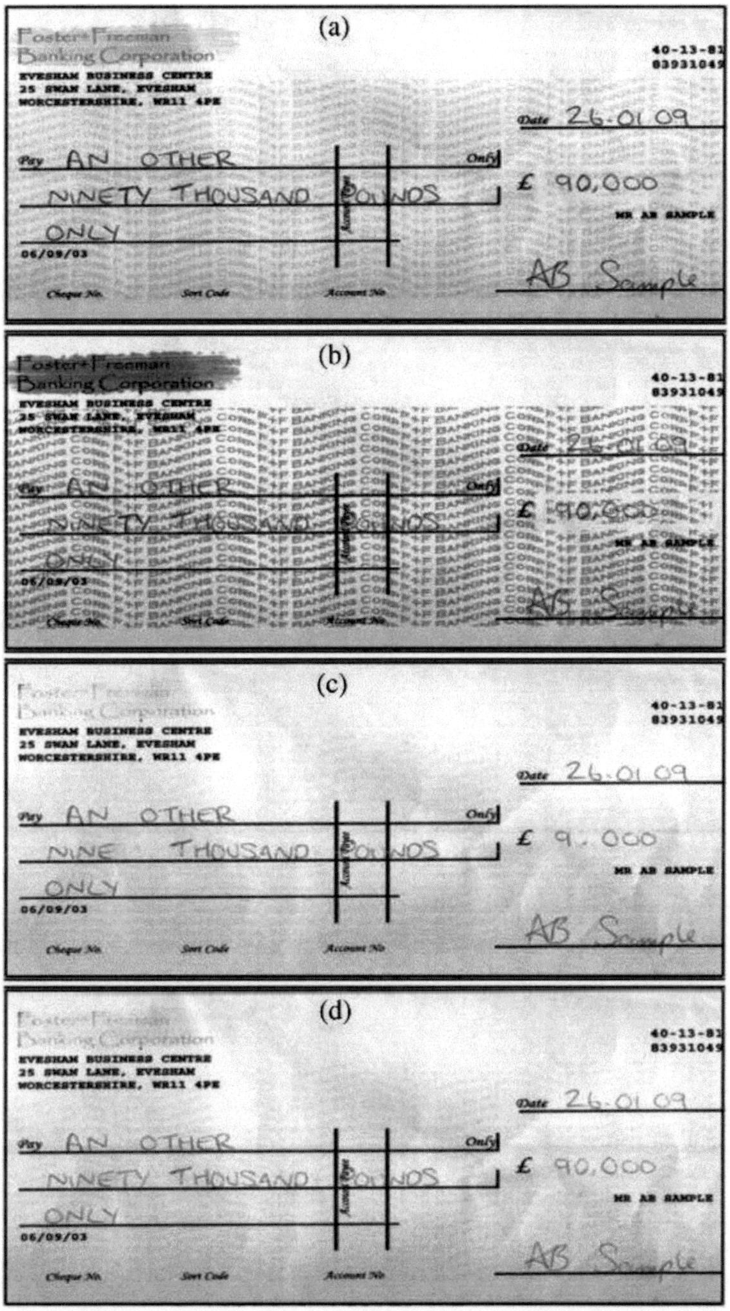

Figure 7. Four of 125 initial pictures used for processing of the HSI, obtained in spectral range 401–998 nm. (a) The picture corresponding to 492,50 nm; (b) 636,69 nm; (c) 780,80 nm; (d) 997,15 nm.

Thus, forgery detection was performed with an interactive user participation, which used information of an indirect nature ("the main text" and "handwritten text different from the main one").

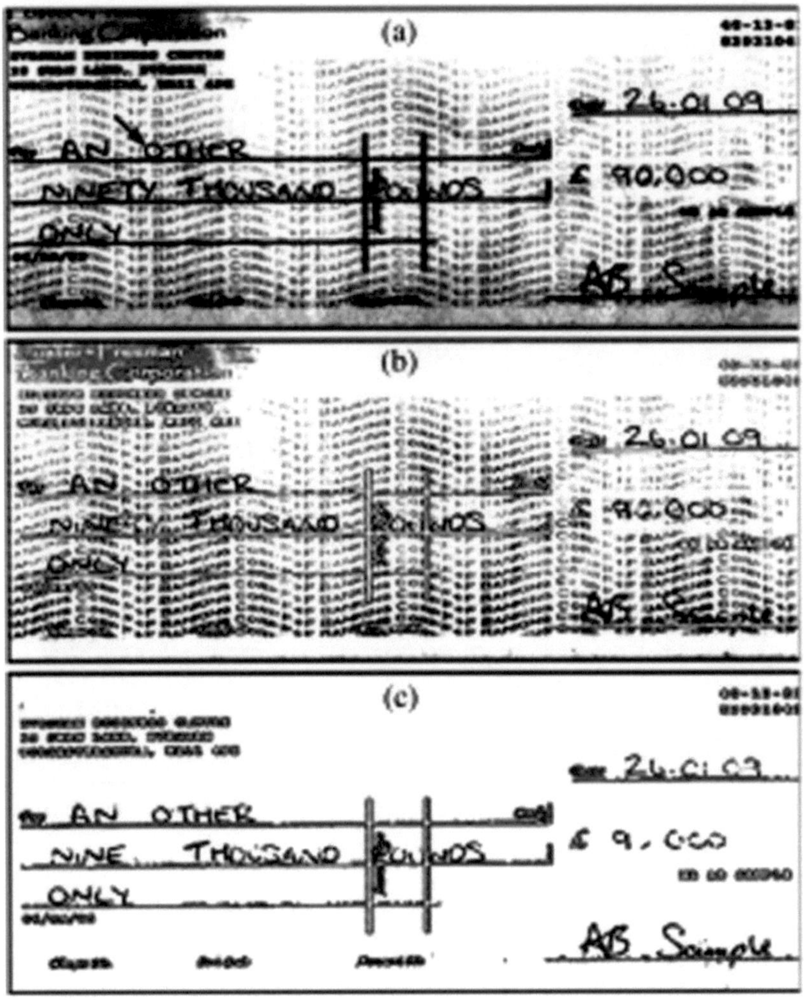

Figure 8. Distinguishing the text whose spectrum is similar to the spectrum of inks of the main handwritten text. (a) The image that corresponds to 545,37 nm with an element of the main text indicated by an arrow; (b) the synthesized image for which the main handwritten text was taken as a reference; (c) the synthesized image with correction of the brightness and contrast.

Detection of objects of interest can be performed with a greater degree of automation by combining automatic segmentation and target visualization. At the first stage, automatic segmentation of the spectra of the hyperspectral image is performed. As a result of automatic segmentation, the main spectra are determined. After such preliminary segmentation, an interactive correction can be made using the target visualization of the measure of similarity with the samples-representatives of the corrected areas.

As an example, we performed HSI processing with the joint use of automatic segmentation and the target visualization of the main spectra [6–8]. The main spectra are determined by automatic segmentation. The visualization of the similarity of the spectra of current pixels to the main spectra is shown by the synthesized images in **Figure 11a**. Interactive correction of these synthesized images, performed with a change in the brightness range, is shown in **Figure 11b**.

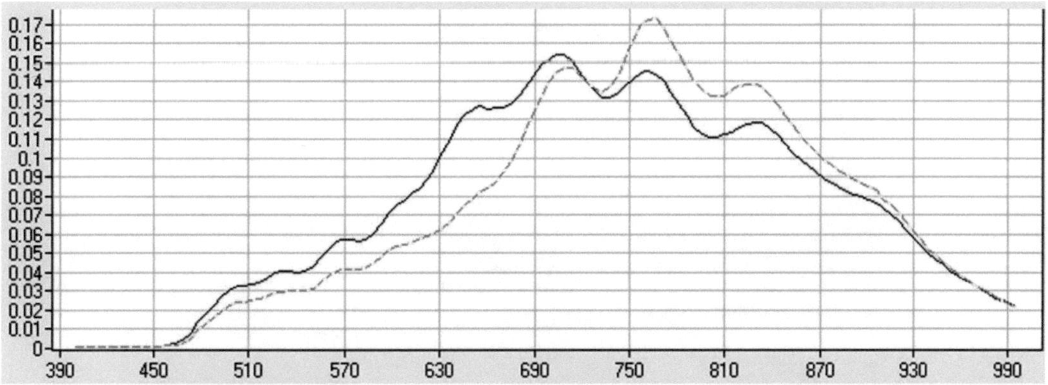

Figure 9. Spectrum of ink of the main text and spectrum of ink of the forgery.

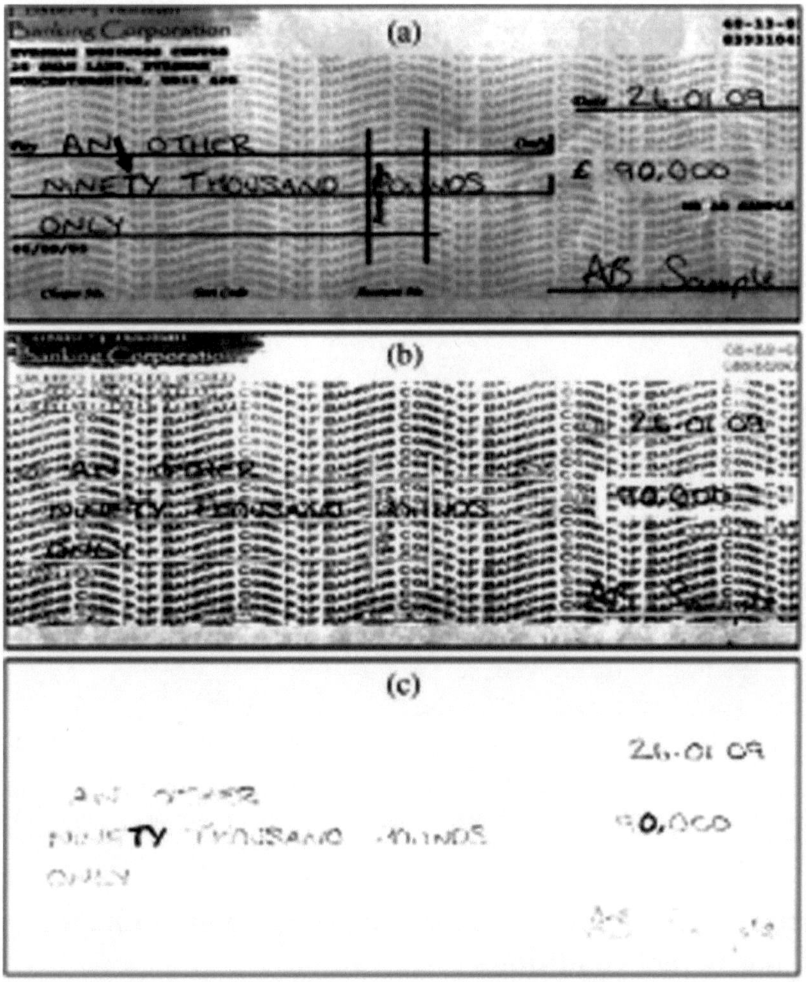

Figure 10. Allocation of the handwritten text of the forgery. (a) The image that corresponds to 545,37 nm with an element of the forgery indicated by an arrow; (b) the synthesized image from the spectrum of the inks of the forgery (the arrow in (a)) and (c) the synthesized image with correction of the brightness and contrast.

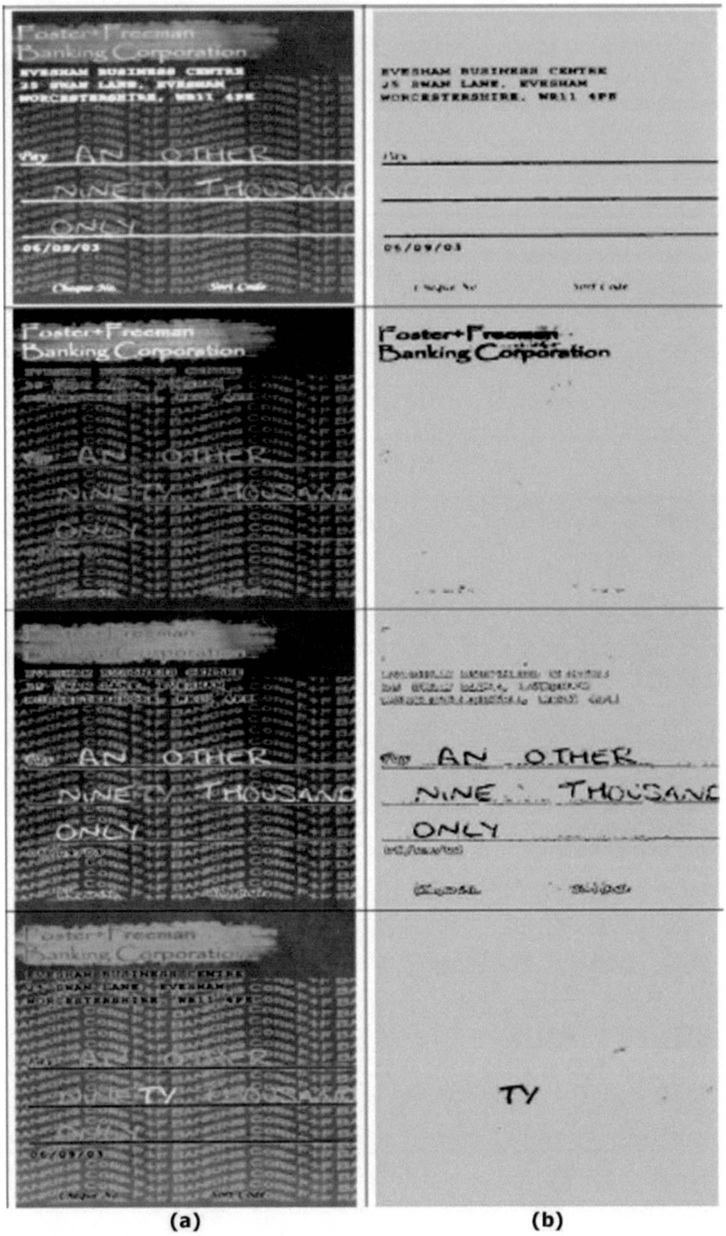

Figure 11. Results of the combined use of automatic segmentation and target visualization of similarity to the main spectra. (a) Visualization of similarity to the spectra obtained by automatic segmentation; (b) the images edited by selecting the brightness range.

5. Automatic high-accurate coordinate superposition of images obtained in different conditions

Efficient interpretation of images of objects and scenes may be often achieved only with combined processing of video data obtained at different time, by different surveying devices, within different spectral ranges. For combined processing of such images automatic high-accurate coordinate superposition of them is required. Such images, even being obtained with the same aspects and at the same scale, considerably differs one from other. Thus usual correlation methods of image superposition may be inadequate [4]. There are two appreciably different approaches to image coordinate superposition: methods based on areas and methods based on image details. Each of these approaches has various advantages and disadvantages. These two approaches may be combined by means of the method of image processing by similarity to a sample [8].

To carry out the correlation coordinate superposition of these images, each of them is suggested to be preliminary transformed by the method of similarity to a sample involving, as an attribute space, textural, gradient features and features of adjacency.

The method does not use reference points of a scene directly. Transforming each of the superposed images by the method of similarity to a sample increase reliability of the superposition methods based on areas. Indeed, this transforming allows us to increase similarity of the images when true superposing of them and to increase their differences when relative shifting. Such transformations can be performed quite a lot, but as much as different samples exist in the fixed image.

For each sample, the correlation method determines the coordinates of the shift. The samples are chosen automatically (without the assistance of user). All of the available samples may be used. But if the images have a lot of samples, they can be selected randomly. For the selected samples a distribution is plotted.

Example of defining a coordinate shift the image obtained in the visible range with the image obtained in the infrared range, is presented in **Figure 12**. Presented in these shots images of the scene in the visible and IR ranges (**Figure 12a** and **b**), have parallel shift. The proposed method determined the shift; the displacement was estimated in fractions of a pixel. **Figure 12c**

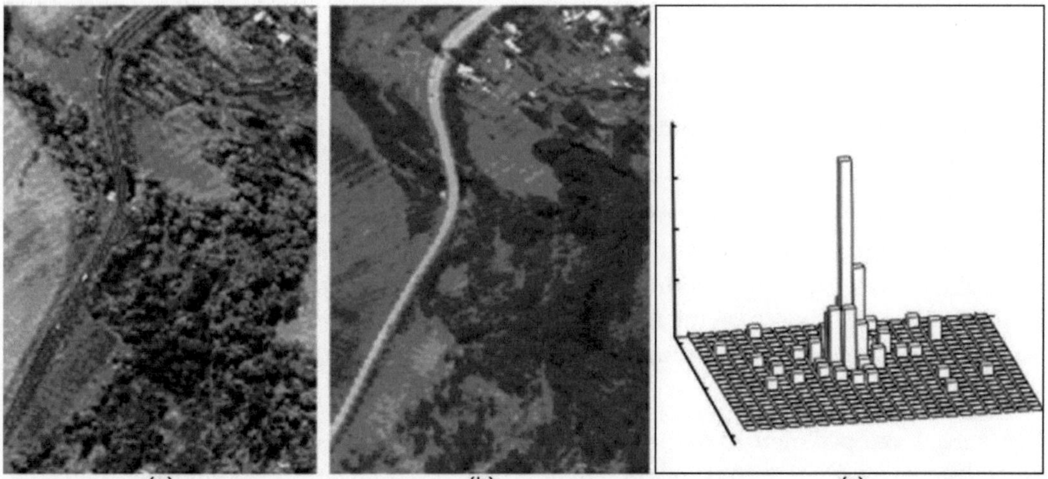

(a) (b) (c)

Figure 12. Determining shift of the images by coordinates. (a) and (b) the initial images; (c) histogram of the shift estimations.

represents distribution of shifts obtained for various samples. Quality of the superposing may be estimated by statistical characteristics. For example, for this distribution of shifts the following characteristics were calculated: mode of the distribution corresponds to shifts: −0.25 pixels along OX axis, −0.5 pixels along OY axis; the mean shift along OX axis is equal to −0.2 pixels and the mean shift along OY axis is equal to −0.6 pixels; the standard deviation along OX axis is equal to 0.52 pixels and the standard deviation along OY axis is equal to 0.55 pixels.

6. The target visualization of hyperspectral images in the sphere of environmental remote sensing

Up-to-date equipment allows you to collect and analyze information in very narrow spectral bands and is used in various applications. In particular, for water monitoring, hyperspectral technology makes it possible to identify the distribution of algae by the concentrations of chlorophyll, and also to detect impurities of mineral substances in water and to determine the coefficients of cloudiness. Hyperspectral data can also be used to identify threatened vegetation types, some chemical toxins, changes in the chemical composition of soils, etc.

Processing hyperspectral images in remote sensing is related with the difficulties arising from the need to take into account the various distortions that appear in the process of formation of the HSI.

In known methods of processing remote sensing data before using hyperspectral information, the data pre-processing is performed. The main stages, of the pre-processing are:

- radiometric correction (correcting the distortion due to the uneven sensitivity of the elements of the detectors, accounting for the effects of the atmosphere),
- geometric correction (georeferencing);
- combining images obtained in different spectral bands,
- recalculation of data of distant shooting in values of coefficients of spectral brightness [10].

Such pretreatment is complex and requires additional information, time and computational resources.

Using the target visualization allows us to facilitate the processing of remote sensing data. Really, the target visualization allows us to significantly reduce influence on results of decryption of conditions of shooting (a condition of the atmosphere, exposition parameters), as a sample can be chosen on the initial images. In this case, the values of attributes of a the sample are obtained under the same conditions of shooting, as the values of attributes of the visualized scene.

Using the developed method of the target visualization the work on creation of the information layer "Swamp" on the territory of St.-Petersburg has been done [9]. Remote sensing data, 435 micro landscapes of 11 types on total area of about 40 sq. km were mapped.

Majority of the swamps in the city is under strong anthropogenic influence due to which the images of the wetlands on the territory of St.-Petersburg are very diverse (by type) and variegated (by the spectral and spatial characteristics) structure, which creates a big problem in their interpretation by space images. As input data, the following materials were used:

- the multispectral images of the territory of St. Petersburg obtained by the Quickbird satellite equipment with the resolution 2.4 m on per pixel;
- the materials of field researches;
- the available landscape and thematic maps.

Here, the images of the territory of town Lomonosov (near by the seaport) in four spectral bands, obtained in August by means of the Quickbird satellite equipment, are presented (**Figure 13**).

Using the linked archival map and the results of field researches, the necessary samples area on the image was allocated. As the sample area in this case, understand the territory which characteristics precisely match the description of this type of a micro landscape in the accepted scheme of classification.

For an example the polygon corresponding to the coastal open swamp was chosen. Due to the complexity of the research object, additional information was used in the form of additional images (**Figure 14**):

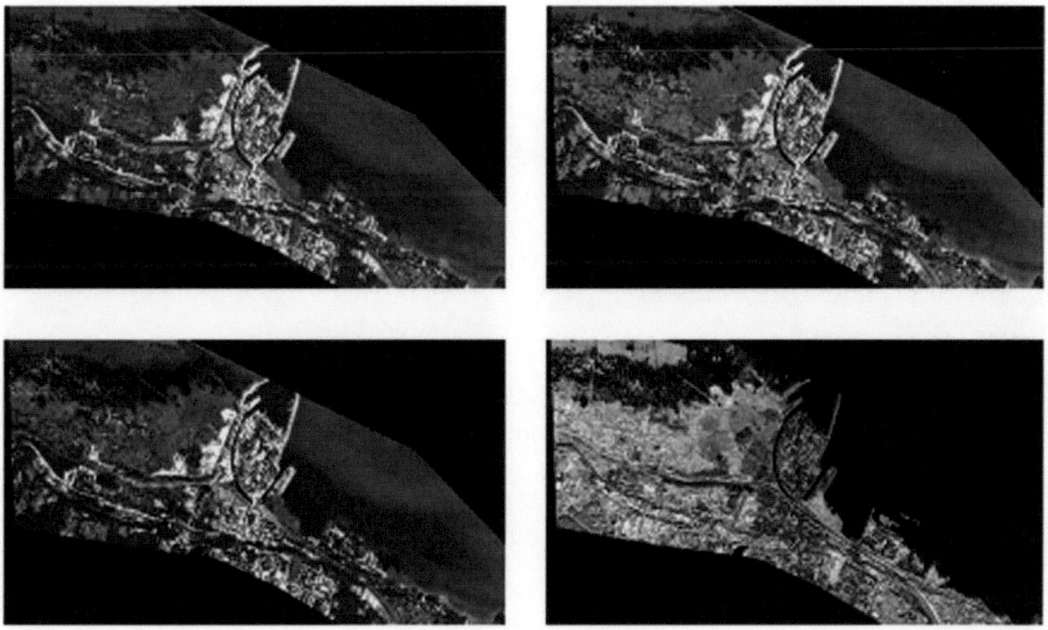

Figure 13. Images of the territory of town Lomonosov (near by the seaport) in four spectral bands obtained in August by means of the Quickbird satellite equipment.

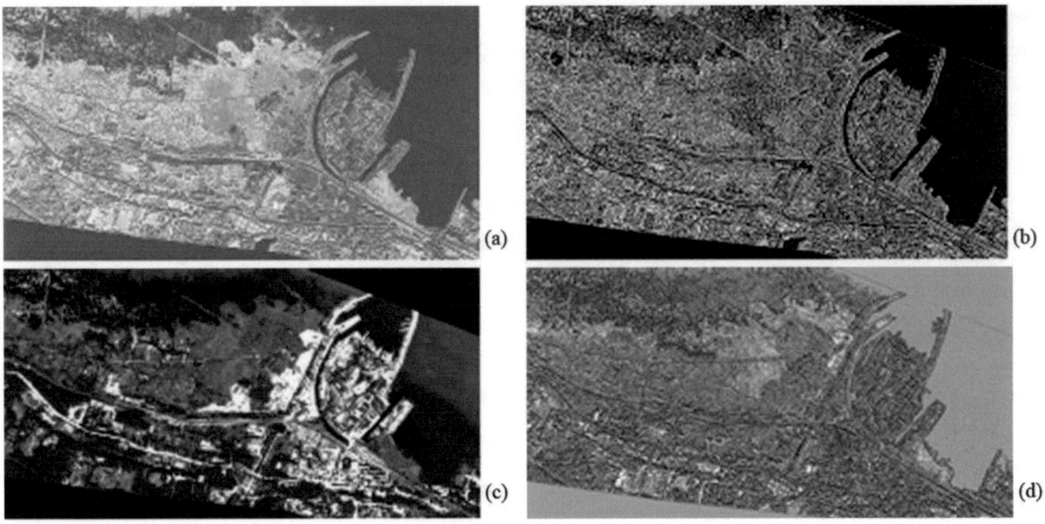

Figure 14. The additional images derived from the original images by using different filters: (a) high pass filter; (b) Laplas filter; (c) local sigma filter; (d) visualization of the texture of one of the original images.

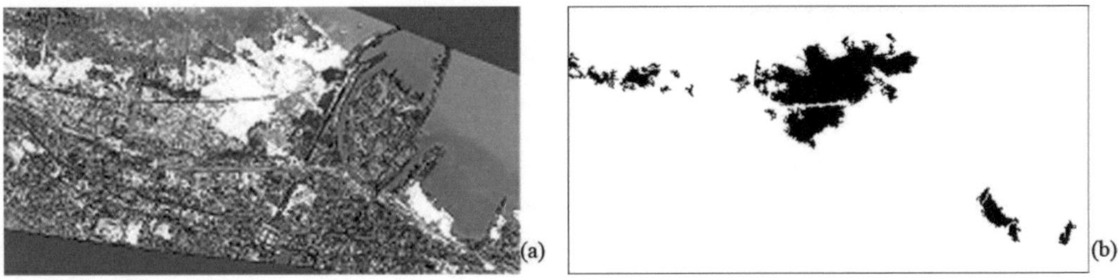

Figure 15. The result of the target visualization of the coastal open swamps: (a) the result of the target visualization; (b) the result of the interactive correction.

- images derived on the base of spatial-spectral, textural, gradient and morphological characteristics of the original images;
- images obtained at other season.

The result of the target visualization the coastal open swamps is presented in **Figure 15a**. The result of the interactive correction of the brightness and contrast are presented in **Figure 15b**.

7. Search for the boundaries of homogeneous regions on hyperspectral image

The target visualization method is also applicable to the problem of finding out the boundaries of homogeneous regions on hyperspectral image.

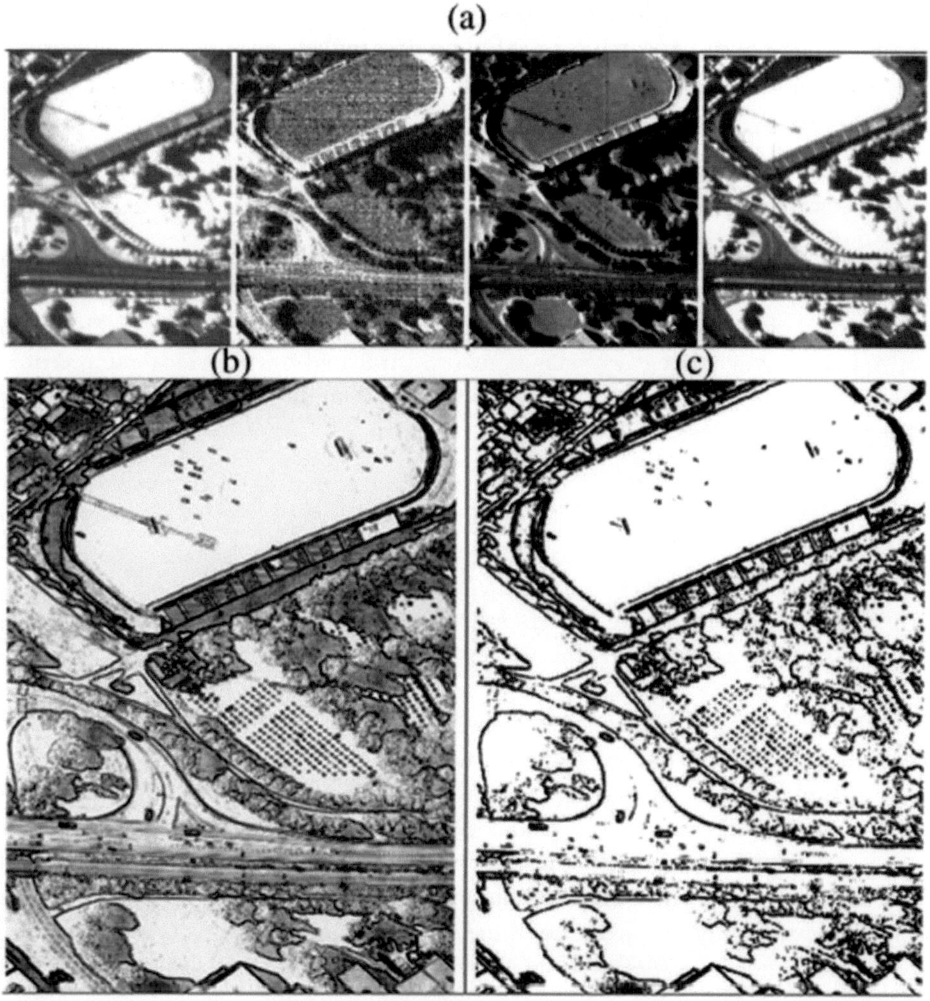

Figure 16. Detection of boundaries of homogenous regions for hyperspectral image. (a) Some selection of pictures from the hyperspectral cube; (b) the synthesized image which visualizes the maximum value of spectral angles within local regions of 3 × 3 pixels; (c) the result of allocation the boundaries.

Methods of search the boundaries are well developed for grayscale images [3, 10]. Grayscale image is considered as a function of two variables (x, y) and it is assumed that the boundaries of the regions correspond to maxima of the gradient of this function. The main problem of the search of boundary is instability their results to noise. Due to some redundancy in the joint use of multi-channel data, the selection of boundaries on the hyperspectral image is more stable to noise. The target visualization can be used to find out the boundaries of homogeneous regions. For this purpose, the hyperspectral image is transformed in the following way. For each pixel of the hyperspectral image with coordinates (m, n) the maximum spectral angle in a local region of size 3 × 3 pixels with center at (m, n) is calculated. The hyperspectral image is converted to a single-band image in accordance with the value of the spectral angle ug_{mn}.

To the obtained synthesized image in tones of gray that contains information of all channels, the known methods of search for the boundaries on halftone images can be applied. In addition, visualization of the gradient makes it possible to interactively edit the result of the boundary outlining.

Figure 16 demonstrates this method of selection of boundaries of homogeneous regions on the hyperspectral image of an urban area.

The hyperspectral image was provided by the Norwegian company Norsk Elektro Opitkk (330 channels). **Figure 16a** shows some shots from this hyperspectral cube. Note that the images of the individual channels are fuzzy, and some channels are highly noisy. In **Figure 16b** the synthesized image, visualizing ug_{mn} the maximal spectral angles within local regions of 3 × 3, is presented. The result of the interactive selection of the boundaries on the basis of this synthesized image is shown in **Figure 16c**.

8. Conclusion

Use of the target visualization of hyperspectral images has a number of useful properties. Image processing by using this method is easy and quite intelligible to applied specialists. Visual data representation makes it easy for user to take a decision, allowing him to use unformalized signs according to his experience and to interactively control process of recognition of objects of interest. The transform performs rapid and compact representation of data. Indeed, measure of similarity is computed quickly and is a scalar value, that allows representing the synthesized image in one channel, and allows us to transmit it over one communication channel. It is easy enough to implement both interactive and automatic selection of objects of interest by choosing a threshold for brightness on a synthesized image. The method allows us to considerably diminish the influence of survey conditions (state of atmosphere, exposure parameters) on results of visual interpretation, because a sample can be chosen from the initial images. In this case the spectral attribute values are obtained under the same survey conditions as the feature values of the visualized scene.

The method is implemented in the basic software [10], which allows the target-visualization of hyperspectral images both automatically and with interactive user participation.

The program performs the following operations.

- Shows the source images in each channel.
- Implements a method for visualizing hyperspectral image using the method of similarity to the reference spectrum.
- Changes the brightness and contrast of the visualization result.
- Synthesizes a color image in a pseudocolor using different reference spectra and different similarity measures.

- The program includes tools that allow you to:
- Zoom the image. To do making a selection of region of an arbitrary shape on the image
- Cut an arbitrary fragment of a hyperspectral image and perform the same actions with it as with the whole image.
- Write the results of the processing to the file.
- View the spectral curves in different, user-selected, spatial points of the image.
- To store the spectra measured in the process of operation in the spectral library and use them later.

Author details

Tatiana A. Sheremeteva*, Gennadiy N. Filippov and Andrey M. Malov

*Address all correspondence to: mapa45@mail.ru

Scientific Research Institute of Optoelectronic Instrumentation, Sosnovy Bor, Leningrad Oblast, Russia

References

[1] Sheremetyeva TA. Method of representation of remote sensing data that facilitates visual interpretation. In: Warmbein B, editor. Proceedings of the 5th International Conference on Space Optics (ICSO 2004); 30 March–2 April 2004; Toulouse, France. ESTEC, Noordwijk, The Netherlands: ESA Publication Division; 2004. pp. 199-201

[2] Sheremet'eva TA, Filippov GN. Method of transforming images. Russian Federation Patent No. 2267232. Invention Bulletin. 2005;**12**(36):265

[3] Gonzalez R, Woods R. Digital Image Processing. 2nd ed. Prentice Hall, Upper Saddle River, N.J.: Pearson Education, Inc.; 2002. 1072 pp

[4] Jahne B. Digital Image Processing. 6th ed. Berlin, Heidelberg: Springer-Verlag; 2005. 584 pp

[5] Wiegant CL, Richardson AJ, Escobar DE, et al. Vegetation indices in crop assessments. Remote Sensing of Environment. 1991;**35**:105-119

[6] Sheremet'eva TA, Filippov GN, Malov AM. Using the target-visualization method to process hyperspectral iamges. Opticheskii Zhurnal. 2015;**82**(1):32-36

[7] Sheremet'eva TA, Filippov GN, Malov AM. Software for visualization of hyperspectral images. Certificate of State Registration of Software No. 2013615622. Copyrighter-OAO "RI OED". 2013

[8] Sheremet'eva TA, Filippov GN. Method of determination of spatial shift of image. Russian Federation Patent No. 246013; 2012

[9] Sheremet'eva TA, Osipov VM, Filippov GN. Interpretation of swampy microlandscapes by space images using spectral and textural attributes of sample ground. In: International Symposium of CIS "Atmospheric Radiation and Dynamics (ISARD - 2009); June 22-26; St.-Petersburg, Russia. 2009. pp. 19-20

[10] Sheremet'eva TA, Filippov GN, Malov AM. Method of visualization of hyperspectral images in remote sensing system. Scientific and Technical Bulletin of the S.-Petersburg State University of ITMO. 2007;**7**(43):293-300

Chapter 9

Graph Construction for Hyperspectral Data Unmixing

Zeng Li, Jie Chen and Susanto Rahardja

Additional information is available at the end of the chapter

http://dx.doi.org/10.5772/intechopen.73158

Abstract

This chapter presents graph construction for hyperspectral data and associated unmixing methods based on graph regularization. Graph is a ubiquitous mathematical tool for modeling relations between objects under study. In the context of hyperspectral image analysis, constructing graphs can be useful to relate pixels in order to perform corporative analysis instead of analyzing each pixel individually. In this chapter, we review fundamental elements of graphs and present different ways to construct graphs in both spatial and spectral senses for hyperspectral images. By incorporating a graph regularization, we then formulate a general hyperspectral unmixing problem that can be important for applications such as remote sensing and environment monitoring. Alternating direction method of multipliers (ADMM) is also presented as a generic tool for solving the formulated unmixing problems. Experiments validate the proposed scheme with both synthetic data and real remote sensing data.

Keywords: hyperspectral imaging, graph construction, spectral unmixing, graph regularization, spectral-spatial correlation

1. Introduction

Hyperspectral imaging analysis has found a wide range of applications including agricultural monitoring, environment detection, meteorological information forecast, medical examination, and camouflage tests [1]. In a hyperspectral image, pixels are typically mixtures of several pure material components due to the limitation of spatial resolution and intimate interactions among materials. Spectral unmixing is thus one of the most important tasks in hyperspectral data analysis, aiming to separate the observed pixel spectra into a collection of constituent spectra, or spectral signatures, called endmembers and to estimate fractions associated with each component called abundances. Spectral unmixing provides a comprehensive and quantitative mapping of the elementary materials that are present in the acquired data, and it is widely used for many applications, such as determining the constitutions of geological mixtures and performing a classification of crops and vegetation.

Most spectral unmixing approaches are designed based on pre-assumed mixture models that describe in an analytical way how the endmembers are combined to mixed spectra measured by the sensor [2]. The linear mixing model (LMM) is the most widely used one, and it assumes that the mixing occurs at a macroscopic scale [3]. A measured spectrum is the linear combinations of the endmembers, weighted by the fractional abundances. To be physically interpretable, LMM is usually performed with two physical constraints, abundance nonnegative constraint (ANC) and abundance sum-to-one constraint (ASC). Multiple scattering effects and intimate interactions in real environment require using nonlinear mixture models. Such models include intimate mixture model [4], bilinear model [5], linear-quadratic mixing model (LQM) [6], polynomial post-nonlinear mixing model (PPNM) [7], to cite a few. However, due to the simplicity and interpretability of the analysis results, LMM-based unmixing strategies are mostly used in practice [2]. A number of unmixing algorithms are proposed, including long-standing geometrical and statistical approaches and the recently introduced sparse regression-based unmixing algorithms [8–11].

Considering inherent spatial-spectral duality exists in hyperspectral data, regularized unmixing algorithms have been proposed in recent years to make use of spectral information and spatial contextual information to enhance the unmixing performance. For instance, in [8], authors introduce a total variation (TV) regularizer to promote spatial consistency of estimated abundances. In [12], the quadratic Laplacian regularization is introduced based on graph representation. In [13], authors present a spatial spectral coherence regularization that relates abundance estimation of a pixel to that of its neighboring pixels with spectral similarities. In [14], authors perform the unmixing with low-rank spatial regularization within fixed-size square windows.

However, it is necessary to establish a frame for these various ways of regularization via a proper mathematical tool. A graph is a ubiquitous structure that describes the connection relationship of a set of vertices. Graph theory is actively used in fields such as biochemistry (genomics), electrical engineering (communication networks and coding theory), computer science (algorithms and computation), and operations research (scheduling) [15]. In addition, several works apply graph theoretical techniques to hyperspectral images, including methods for dimensionality reduction [16], anomaly detection [17], and classification [18]. In the context of hyperspectral data unmixing, a graph can be used to model relations of spatial and spectral information of hyperspectral pixels. In this chapter, we will present a variety of ways to construct a graph for hyperspectral unmixing and formulate the associated unmixing problem with solutions given by the alternating direction method of multipliers (ADMM) strategy.

The remainder of the chapter is organized as follows: Section 2 introduces graph theory and graph construction methods in the context of hyperspectral unmixing. Section 3 formulates the sparse linear unmixing problem based on graph regularization. In Section 4, the solution to the formulated problem is derived via the ADMM algorithm. Section 5 reports the experiment's results. Finally, Section 6 concludes this chapter.

2. Graph construction

2.1. Introduction to graphs

We firstly review some fundamental elements of a graph. A graph is a general data structure described by $G = (V, E)$, where a finite set of vertices, also called nodes, is denoted by V and a finite set of pairs of the form $(v_i v_j)$ is referred to as edges. Edges indicate the relation between vertices, and they can be directed or undirected. Directed edges utilize ordered pairs of points that indicate the source and sink of each connection, that is, $(v_i v_j)$ represents an edge from v_i to v_j. Undirected edges only indicate the relationship between vertices and do not consider the ordering, that is, $(v_i v_j)$ is the same as $(v_j v_i)$. We may associate each edge with a weight to describe the importance or the cost of this connection (**Figure 1**).

In a simple setting, if two vertices are connected by an edge, the weight is set to 1, otherwise the weight is 0. The following part introduces some other ways to measure the similarity among vertices, in other words, to define the weights. We can use either adjacency matrices or incidence matrices to describe graphs depending on the type of operations to be performed. Elements of the matrix **A** indicate whether pairs of vertices are connected or not in a graph. Element A_{ij} is 1 when there is an edge from vertex i to vertex j and zero when there is no edge. If the graph is undirected, the adjacency matrix is symmetric. Incidence matrices show the relationship between vertices and edges. An undirected graph can have two kinds of incidence matrices: unoriented and oriented matrices. An oriented incidence matrix in the undirected graph can be denoted by $\mathbf{B} \in \mathbb{R}^{n \times m}$, where n is the number of vertices and m is the number of edges. That is, in the column of edge e_k , the positive undirected graph can be denoted by $\mathbf{B} \in \mathbb{R}^{n \times m}$, where n is the number of vertices and m is the number of edges. That is, in the column of edge e_k , there is positive weight A_{ij} in the row corresponding to one vertex v_i of e_k and negative weight $-A_{ij}$ in the row corresponding to the other vertex v_j of e_k , and all other rows are set to 0.

In addition, a degree matrix for a graph is a diagonal matrix $\mathbf{D} = \text{diag}(d_1, \cdots d_i, \cdots d_n)$, where n is the number of vertices and d_i is the degree of the vertex v_i in G. The degree matrix is indicating every vertex's degree which is the number of edges connecting to one vertex. It is

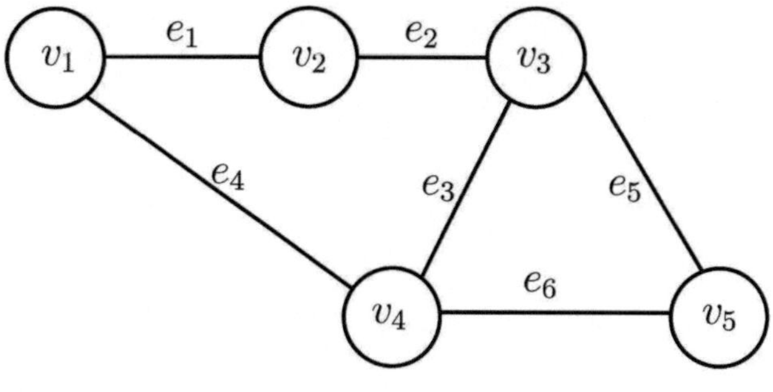

Figure 1. Example of a graph.

normally used together with the adjacency matrix to construct the Laplacian matrix **L** of a graph, which is $\mathbf{L} = \mathbf{D} - \mathbf{A}$.

2.2. Graph construction for hyperspectral images

In this part, we elaborate the ways to construct graphs in the context of hyperspectral image analysis. The performance of spectral unmixing is closely tied to the graph construction of images, and in the hyperspectral remote sensing literature, there are a number of techniques. In [19], authors summarize a survey of spectral graph construction techniques and discuss advantages and disadvantages of these techniques. Generally, each pixel can be viewed as a vertex (or node), and each vertex is associated with a continuous spectrum. A set of edges can be set and assigned with weights in different senses as presented here below.

2.2.1. Four-neighbor graph

A common and straightforward construction is to consider the four-neighbor graph, where every vertex (i.e., every pixel) is connected to four nearest spatially adjacent neighboring pixels, as illustrated in **Figure 2**.

2.2.2. Threshold-compared graph

Another alternative to construct a graph is to calculate all pairwise distances and an edge is placed if the distance between two vertices is less than a user-predefined threshold. The distance in the hyperspectral image can be measured using the spectral distance or spatial distance. For instance, $\boldsymbol{v}_i$ and $\boldsymbol{v}_j$ are two vertices that are associated with spectral vectors with L bands, then their Euclidean distance is $\|\boldsymbol{v}_i - \boldsymbol{v}_j\| = \sqrt{\sum_{k=1}^{L} (v_{ik} - v_{jk})^2}$.

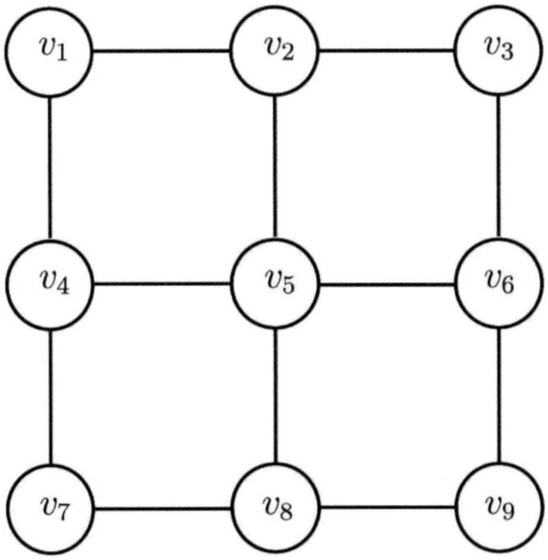

Figure 2. Spatial four-neighbor method.

2.2.3. K-nearest neighbor graph

Constructing a graph with *k*-nearest neighbors (k-NN) is a popular method. In this case, an edge is set between two vertices if vertex v_j is in *k*-NN of vertex v_i. Each vertex has its own *k*-nearest neighbors. Consequently, the graph is a directed graph. It is worth noting that constructing such a graph requires calculating all pairwise distances and ordering these values on each vertex, and these operations lead to high computational costs.

2.2.4. Spatial-spectral graph

As pixels in a hyperspectral image possess spatial locations and spectral signatures, it can be beneficial to construct a graph by incorporating both spatial and spectral information. For instance, a graph can be constructed with local four neighborhood pixels and by considering spectral similarity among pixels, as described in **Figure 3**.

2.2.5. Weighted graph

Above methods construct unweighted graphs with only connection indications among pixels. Several other methods further impose weights on each edge. For instance, spectral similarity measured by the Gaussian kernel can be used to define weights:

$$A_{ij} = \exp\left(-\frac{\|v_i - v_j\|^2}{2\sigma^2}\right) \tag{1}$$

where σ is the kernel bandwidth and defined by users. As a generalization, a radial basis function (also called a diffusion kernel) in spectral distance with two parameters σ_i and σ_j is introduced in [20], given by:

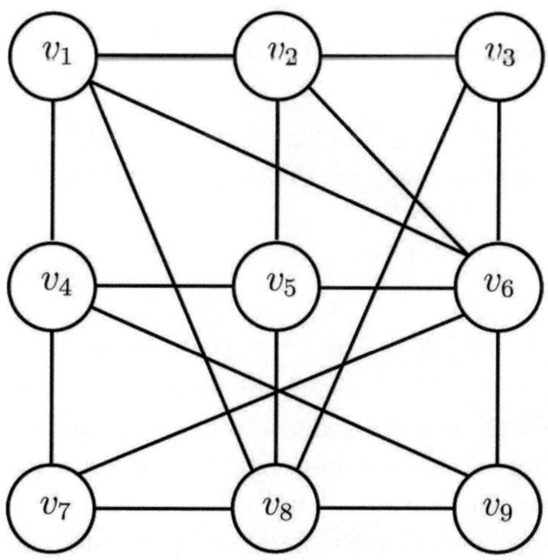

Figure 3. An example of four spatial neighbors and *k*-NN spectral neighbors.

$$A_{ij} = \exp\left(-\frac{\|v_i - v_j\|^2}{\sigma_i\sigma_j}\right). \tag{2}$$

Weights can also be calculated by considering both spatial and spectral information. For instance, [21] proposes to define weights by:

$$A_{ij} = \exp\left(-\frac{\|v_i - v_j\|^2}{c_{ij}\sigma_i\sigma_j}\right)\cdot\exp\left(-\frac{\|x_i - x_j\|^2}{\sigma_d^2}\right) \tag{3}$$

where x_i is the spatial coordinates of pixel v_i , c_{ij} is an integer indicating the number of common neighbors between v_i and v_j , σ_i and σ_j are defined in [20], and the parameter σ_d is defined by users which limits the size of regions spatially. In [22], authors consider the similarity of the spectral angle instead of the spectral Euclidean distance.

$$A_{ij} = \exp\left(-\theta_{ij}\right)\cdot\exp\left(-\frac{\|x_i - x_j\|^2}{\sigma}\right) \tag{4}$$

where θ_{ij} denotes the spectral angle between v_i and v_j , x_i is the spatial coordinates of pixel v_i and σ is the parameter defined by users. Note that some schemes of calculating weights can make edges to be severed so as to change the structure of the graph [19].

There are also some other methods to construct graphs adapted to hyperspectral images, such as adaptive nearest neighbor graphs, density-weighted *k*-NN graphs, and shared nearest-neighbor graphs [19].

3. Graph-based regularization in unmixing

With a constructed graph at hand to model the relation of pixels, in this section, we present the way to perform a sparse unmixing with the graph regularization.

3.1. Sparse unmixing

Consider the linear mixing model: $\mathbf{y} = \mathbf{S}\mathbf{x} + \mathbf{n}$, where $\mathbf{y} \in \mathbb{R}^L$ is one observed pixel with L spectral bands, $\mathbf{S} = [\mathbf{s}_1, \mathbf{s}_2, \cdots, \mathbf{s}_R] \in \mathbb{R}^{L\times R}$ is the library of spectral signatures including R pure spectral signatures, and $\mathbf{x} \in \mathbb{R}^R$ is an abundance vector, $\mathbf{n}$ is the additive white noise vector. Since it is often that an observed pixel is only composed of a small number of materials in the library, the majority of entries of the abundance vector $\mathbf{x}$ are zero-valued, namely, $\mathbf{x}$ is sparse. Assuming the library is available beforehand and the spare unmixing problem can be defined as [23]:

$$\min_{\mathbf{x}} \ \frac{1}{2}\|\mathbf{y}-\mathbf{S}\mathbf{x}\|_2^2+\lambda\|\mathbf{x}\|_1 \quad \text{subject to}: \mathbf{x}\geq 0 \tag{5}$$

where λ is the regularized parameter.

In this chapter, we formulate the problem without ASC constraint because of using the l_1 norm regularization. Moreover, the validity of ASC is often questioned in the literature for practical scenarios. In what follows, we introduce the graph regularization to the above formulated problem.

3.2. Graph regularization for sparse unmixing

Since a graph relates the pixels of image via spatial and spectral relations, we can regularize the unmixing problem with pixel relations defined by the graph. Let $\mathbf{Y}=[\mathbf{y}_1,\mathbf{y}_2,\cdots\mathbf{y}_n]\in\mathbb{R}^{L\times n}$ be a spectral matrix, where each column is one observed pixel including L spectral bands and n is the number of pixels, and in a graph, n is also the number of vertices. Let $\mathbf{X}=[\mathbf{x}_1,\mathbf{x}_2,\cdots\mathbf{x}_n]\in\mathbb{R}^{R\times n}$ be an abundance matrix in which each column is an abundance vector associated with one observed pixel. With the graph representation of hyperspectral data, we achieve the sparse unmixing by solving the following optimization problem:

$$\min_{\mathbf{X}} \ \frac{1}{2}\|\mathbf{Y}-\mathbf{S}\mathbf{X}\|_F^2+\mu\|\mathbf{X}\|_{1,1}+\lambda_g g_1(\mathbf{X}) \quad \text{subject to}: \ \mathbf{X}\geq 0 \tag{6}$$

where

$$g_1(\mathbf{X})=\sum_{i=1}^{n}\sum_{j=1}^{n}A_{ij}\|\mathbf{x}_i-\mathbf{x}_j\|_1 \tag{7}$$

This graph regularization term Eq. (7) is based on the assumption that if two vertices are connected by an edge, then the abundances of the two vertices are similar. This term measures the differences between all pairs of abundances weighted by their degrees of similarity in the graph. The graph regularization then promotes piecewise constant transitions of estimates among the related pixels. Parameter λ_g controls the regularization strength. Note that we can rewrite Eq. (7) with the incidence matrix $\mathbf{B}$ as:

$$\sum_{i=1}^{n}\sum_{j=1}^{n}A_{ij}\|\mathbf{x}_i-\mathbf{x}_j\|_1=\|\mathbf{X}\mathbf{B}\|_{1,1} \tag{8}$$

Problem Eq. (6) is equivalently expressed as:

$$\min_{\mathbf{X}} \ \frac{1}{2}\|\mathbf{Y}-\mathbf{S}\mathbf{X}\|_F^2+\mu\|\mathbf{X}\|_{1,1}+\lambda_g\|\mathbf{X}\mathbf{B}\|_{1,1} \quad \text{subject to}: \ \mathbf{X}\geq 0 \tag{9}$$

If we use a spatial four-neighborhood graph in this unmixing problem with the weights being simply set to 1 and 0, it can generally be identical with the SUnSAL-TV algorithm [8]. The right term can promote piecewise constant transitions in the fractional abundance among neighborhood pixels and achieve spatial consistency of estimated abundances.

Instead of considering promoting the similarities among estimated abundances, an alternative way is to promote the similarities of reconstructed spectra among the connected pixels. In [24], authors propose a nonlocal TV regularization, with the regularization term given as:

$$g_2(\mathbf{X}) = \sum_{i=1}^{n}\sum_{j=1}^{n} A_{ij}\left\|\mathbf{S}\mathbf{x}_i - \mathbf{S}\mathbf{x}_j\right\|_1 \tag{10}$$

This can also be written with incidence matrix **B** as:

$$g_3(\mathbf{X}) = \|(\mathbf{S}\mathbf{X})\mathbf{B}\|_1 \tag{11}$$

4. Solution to the formulated problem

We propose to solve the formulated unmixing problem Eq. (9) via the ADMM algorithm. In this section, we first briefly review the ADMM algorithm and then apply it to our unmixing problem.

4.1. Introduction of ADMM

ADMM is an algorithm that is intended to blend the decomposability of dual ascent with the superior convergence properties of the method of multipliers. The algorithm solves problems in the form [25]:

$$\begin{aligned} &\min f(x) + g(z) \\ &\text{s.t.}\ \ Ax + Bz = c \end{aligned} \tag{12}$$

with variables $x \in \mathbb{R}^n$ and $z \in \mathbb{R}^m$, where $A \in \mathbb{R}^{p\times n}$, $B \in \mathbb{R}^{p\times m}$, and $c \in \mathbb{R}^p$.

The first step is to write the augmented Lagrangian of problem Eq. (12):

$$L_\rho(x,z,y) = f(x) + g(z) + y^T(Ax + Bz - c) + \frac{\rho}{2}\|Ax + Bz - c\|_2^2 \tag{13}$$

ADMM suggests achieving the optimum via the following iterations:

$$x^{k+1} = \underset{x}{\operatorname{argmin}}\, L_\rho\left(x, z^k, y^k\right) \tag{14}$$

$$z^{k+1} = \underset{z}{\operatorname{argmin}}\, L_\rho\left(x^{k+1}, z, y^k\right) \tag{15}$$

$$y^{k+1} = y^k + \rho\left(Ax^{k+1} + Bz^{k+1} - c\right) \tag{16}$$

where $\rho > 0$. The algorithm is very similar to dual ascent and the method of multipliers: it consists of an x-minimization step Eq. (14), a z-minimization step Eq. (15), and a dual variable update Eq. (16). As in the method of multipliers, the dual variable update uses a step size equal to the augmented Lagrangian parameter ρ.

4.2. Solutions via ADMM

In order to apply the canonical ADMM algorithm to the problem (9), we introduce the auxiliary variables and transform the problem as follows:

$$\begin{aligned} \underset{\mathbf{X}, \mathbf{V}_1, \mathbf{V}_2, \mathbf{Z}}{\text{minimize}} \ & \frac{1}{2}\|\mathbf{Y}-\mathbf{S}\mathbf{X}\|_F^2 + \mu\|\mathbf{V}_2\|_{1,1} + \lambda_g\|\mathbf{V}_4\|_{1,1} + l_{\mathbb{R}_+^{R\times n}}(\mathbf{V}_2) \\ \text{subject to} \ & \mathbf{V}_1 = \mathbf{S}\mathbf{X} \\ & \mathbf{V}_2 = \mathbf{X} \\ & \mathbf{V}_3 = \mathbf{X} \\ & \mathbf{V}_4 = \mathbf{V}_3\mathbf{B} \end{aligned} \tag{17}$$

where l_S is the indicator function of the set S , such as $l_S(\mathbf{x}) = 0$ if $\mathbf{x} \in S$ and $l_S(\mathbf{x}) = +\infty$ if $\mathbf{x} \notin S$. Thus the augmented Lagrangian for Eq. (17) is as follows:

$$\begin{aligned} L(\mathbf{X}, \mathbf{V}_{1\to4}, \mathbf{D}_{1\to4}) = & \frac{1}{2}\|\mathbf{Y}-\mathbf{V}_1\|_F^2 + \mu\|\mathbf{V}_2\|_{1,1} + \lambda_g\|\mathbf{V}_4\|_{1,1} + l_{\mathbb{R}_+^{R\times n}}(\mathbf{V}_2) \\ & + \frac{\rho}{2}\|\mathbf{S}\mathbf{X}-\mathbf{V}_1-\mathbf{D}_1\|_F^2 + \frac{\rho}{2}\|\mathbf{X}-\mathbf{V}_2-\mathbf{D}_2\|_F^2 \\ & + \frac{\rho}{2}\|\mathbf{X}-\mathbf{V}_3-\mathbf{D}_3\|_F^2 + \frac{\rho}{2}\|\mathbf{V}_3\mathbf{B}-\mathbf{V}_4-\mathbf{D}_4\|_F^2 \end{aligned} \tag{18}$$

where $\mathbf{D}_1, \mathbf{D}_2, \mathbf{D}_3, \mathbf{D}_4$ are Lagrange multipliers and ρ is the penalty parameter.

The algorithm steps are as follows:

Step 1. Input the observed pixels $\mathbf{Y}$ matrix, the library $\mathbf{S}$, and the regularization parameters μ, λ_g;

Step 2. Initialization: $\mathbf{X}^{(0)}, \mathbf{V}_1^{(0)}, \cdots, \mathbf{V}_4^{(0)}, \mathbf{D}_1^{(0)}, \cdots, \mathbf{D}_4^{(0)}, \rho$, set $k = 0$

Step 3. Repeat:

Step 4. $\mathbf{X}^{(k+1)} \leftarrow \text{argmin}_{\mathbf{X}} L_\rho\left(\mathbf{X}, \mathbf{V}_1^{(k)}, \mathbf{V}_2^{(k)}, \mathbf{V}_3^{(k)}, \mathbf{V}_4^{(k)}, \mathbf{D}_1^{(k)}, \mathbf{D}_2^{(k)}, \mathbf{D}_3^{(k)}, \mathbf{D}_4^{(k)}\right)$

Step 5. $\mathbf{V}_1^{(k+1)} \leftarrow \text{argmin}_{\mathbf{V}_1} L_\rho\left(\mathbf{X}^{(k+1)}, \mathbf{V}_1, \mathbf{V}_2^{(k)}, \mathbf{V}_3^{(k)}, \mathbf{V}_4^{(k)}, \mathbf{D}_1^{(k)}, \mathbf{D}_2^{(k)}, \mathbf{D}_3^{(k)}, \mathbf{D}_4^{(k)}\right)$

Step 6. $\mathbf{V}_2^{(k+1)} \leftarrow \text{argmin}_{\mathbf{V}_2} L_\rho\left(\mathbf{X}^{(k+1)}, \mathbf{V}_1^{(k+1)}, \mathbf{V}_2, \mathbf{V}_3^{(k)}, \mathbf{V}_4^{(k)}, \mathbf{D}_1^{(k)}, \mathbf{D}_2^{(k)}, \mathbf{D}_3^{(k)}, \mathbf{D}_4^{(k)}\right)$

Step 7. $\mathbf{V}_3^{(k+1)} \leftarrow \text{argmin}_{\mathbf{V}_3} L_\rho\left(\mathbf{X}^{(k+1)}, \mathbf{V}_1^{(k+1)}, \mathbf{V}_2^{(k+1)}, \mathbf{V}_3, \mathbf{V}_4^{(k)}, \mathbf{D}_1^{(k)}, \mathbf{D}_2^{(k)}, \mathbf{D}_3^{(k)}, \mathbf{D}_4^{(k)}\right)$

Step 8. $\mathbf{V}_4^{(k+1)} \leftarrow \text{argmin}_{\mathbf{V}_4} L_\rho\left(\mathbf{X}^{(k+1)}, \mathbf{V}_1^{(k+1)}, \mathbf{V}_2^{(k+1)}, \mathbf{V}_3^{(\mathrm{k}+1)}, \mathbf{V}_4, \mathbf{D}_1^{(k)}, \mathbf{D}_2^{(k)}, \mathbf{D}_3^{(k)}, \mathbf{D}_4^{(k)}\right)$

Step 9. Update the Lagrangian multipliers:

$$\begin{aligned} \mathbf{D}_1^{(k+1)} &\leftarrow \mathbf{D}_1^{(k)} - \left(\mathbf{S}\mathbf{X}^{(k+1)} - \mathbf{V}_1^{(k+1)}\right) \\ \mathbf{D}_2^{(k+1)} &\leftarrow \mathbf{D}_2^{(k)} - \left(\mathbf{X}^{(k+1)} - \mathbf{V}_2^{(k+1)}\right) \\ \mathbf{D}_3^{(k+1)} &\leftarrow \mathbf{D}_3^{(k)} - \left(\mathbf{X}^{(k+1)} - \mathbf{V}_3^{(k+1)}\right) \\ \mathbf{D}_4^{(k+1)} &\leftarrow \mathbf{D}_4^{(k)} - \left(\mathbf{V}_3^{(k+1)}\mathbf{B} - \mathbf{V}_4^{(k+1)}\right) \end{aligned}$$

Step 10. Until stopping criterion is satisfied.

In step 4 of minimizing the augmented Lagrangian with respect to **X**, the solution is:

$$\mathbf{X} \leftarrow \left(\mathbf{S}^T\mathbf{S} + 2\mathbf{I}\right)^{-1}\left(\mathbf{S}^T(\mathbf{V}_1 + \mathbf{D}_1) + \mathbf{V}_2 + \mathbf{D}_2 + \mathbf{V}_3 + \mathbf{D}_3\right) \tag{19}$$

Similarly, the solution of $\mathbf{V}_1$ minimization step 5 is:

$$\mathbf{V}_1 \leftarrow \frac{1}{1+\rho}\left(\mathbf{Y} + \rho(\mathbf{S}\mathbf{X} - \mathbf{D}_1)\right) \tag{20}$$

To compute $\mathbf{V}_2$ in step 6, the solution is the well-known soft threshold [17]:

$$\mathbf{v}_2 \leftarrow \max\left(soft\left(\mathbf{x} - \mathbf{d}_2, \frac{\mu}{\rho}\right), 0\right) \tag{21}$$

where $\mathbf{v}_2$, $\mathbf{x}$, $\mathbf{d}_2$ is the row of $\mathbf{V}_2$, $\mathbf{X}$, $\mathbf{D}_2$, respectively.

The solution of $\mathbf{V}_3$ minimization step 7 is:

$$\mathbf{V}_3 \leftarrow \left(\mathbf{X} - \mathbf{D}_3 + (\mathbf{V}_4 + \mathbf{D}_4)\mathbf{B}^T\right)\left(\mathbf{I} + \mathbf{B}\mathbf{B}^T\right)^{-1} \tag{22}$$

The solution of $\mathbf{V}_4$ minimization step 8 is:

$$\mathbf{v}_4 \leftarrow soft\left(\mathbf{f} - \mathbf{d}_4, \frac{\lambda}{\rho}\right) \tag{23}$$

where $\mathbf{v}_4$, $\mathbf{f}$, $\mathbf{d}_4$ is the row of $\mathbf{V}_4$, $\mathbf{F} = \mathbf{V}_3 \times \mathbf{B}$, $\mathbf{D}_4$, respectively.

5. Experiments

In this section, we illustrate the experimental results via a synthetic hyperspectral data set (denoted by Data 1) and a real hyperspectral data set (denoted by Data 2) with various ways of graph construction.

5.1. Experiments with simulated data sets

In this part, the synthetic data consists of 75×75 pixels and is generated by 9 endmembers. The endmembers are randomly selected from the spectral library advanced spaceborne thermal emission and reflection radiometer (ASTER). Each signature of endmembers has reflectance values measured over 420 spectral bands. The pure regions and mixed regions involved between two and five endmembers, spatially distributed in the form of square regions. The background is a mixture of the five endmembers with the abundance values $[0, 0, 0, 0, 0.1149, 0.0741, 0.2003, 0.2055, 0.4051]$.

The quality of unmixing results for the simulated data can be measured by comparing the estimated and actual abundances using the root mean square error (RMSE),

	15 dB	20 dB	30 dB
Four-neighbor graph	0.0246 $\mu = 0.005, \lambda = 0.05$	0.0184 $\mu = 0.005, \lambda = 0.05$	0.0051 $\mu = 5 \times 10^{\wedge}(-4), \lambda = 0.01$
Spectral-spatial combined graph	0.0085 k = 25; $\mu = 5 \times 10^{\wedge}(-4), \lambda = 0.01$	0.0052 k = 25 $\mu = 0.005, \lambda = 0.01$	0.0021 k = 25 $\mu = 5 \times 10^{\wedge}(-4), \lambda = 0.005$
Threshold-compared graph	0.0025 threshold = 9 $\mu = 5 \times 10^{\wedge}(-4), \lambda = 0.1$	0.0015 threshold = 3 $\mu = 5 \times 10^{\wedge}(-4), \lambda = 0.5$	0.0009 threshold = 0.25 $\mu = 5 \times 10^{\wedge}(-4), \lambda = 0.1$

Table 1. RMSE evaluating performances with different values of SNR, with three constructed graphs and optimal regularized parameters. The values of threshold and k are also shown.

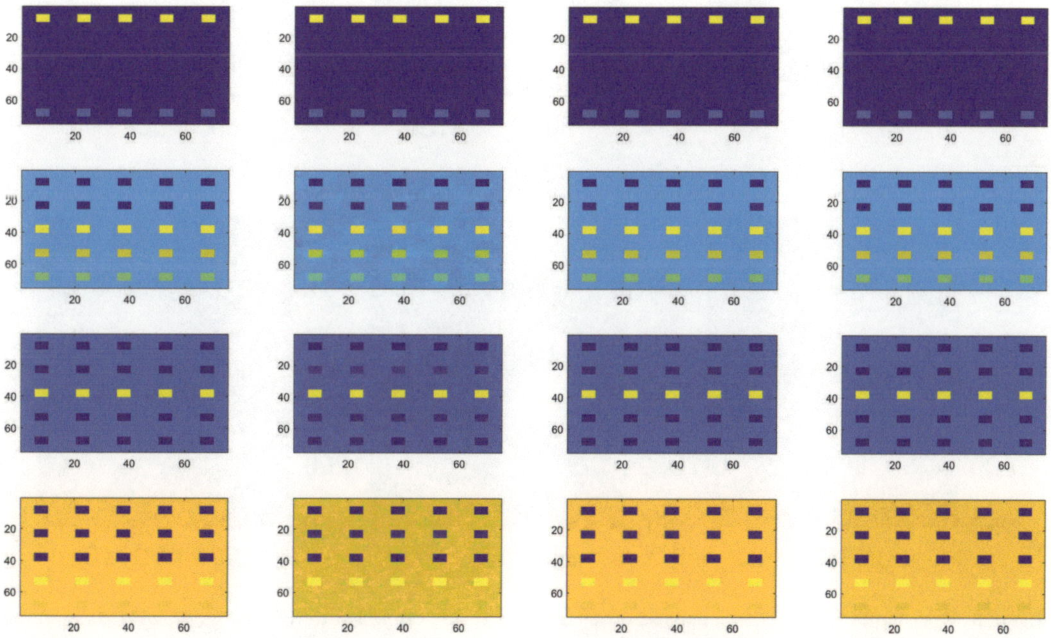

Figure 4. From top to bottom: The abundance maps of first, fifth, sixth, and eighth . From left to right: Real abundance maps, estimated abundance maps with four-neighbor graph, spectral-spatial graph and threshold-compared graph, respectively.

$$RMSE = \sqrt{\frac{1}{nR}\sum_{i=1}^{n} \|\mathbf{x}_i - \widehat{\mathbf{x}}_i\|^2} \tag{24}$$

where $\mathbf{x}_i$ and $\widehat{\mathbf{x}}_i$ are the actual and estimated abundance vectors of the ith pixel, n is the number of pixels, and R is the number of endmembers.

We define the graph based on the simulated data using three methods: the four-neighbor graph, the threshold-compared graph and the spectral-spatial graph respectively.

In the experiment, the threshold-compared undirected graph is constructed as follows:

$$\mathbf{A}_{ij} = \begin{cases} 1 & \text{if} \left\| \mathbf{y}_i - \mathbf{y}_j \right\|_2^2 < threshold \\ 0 & \text{otherwise} \end{cases} \tag{25}$$

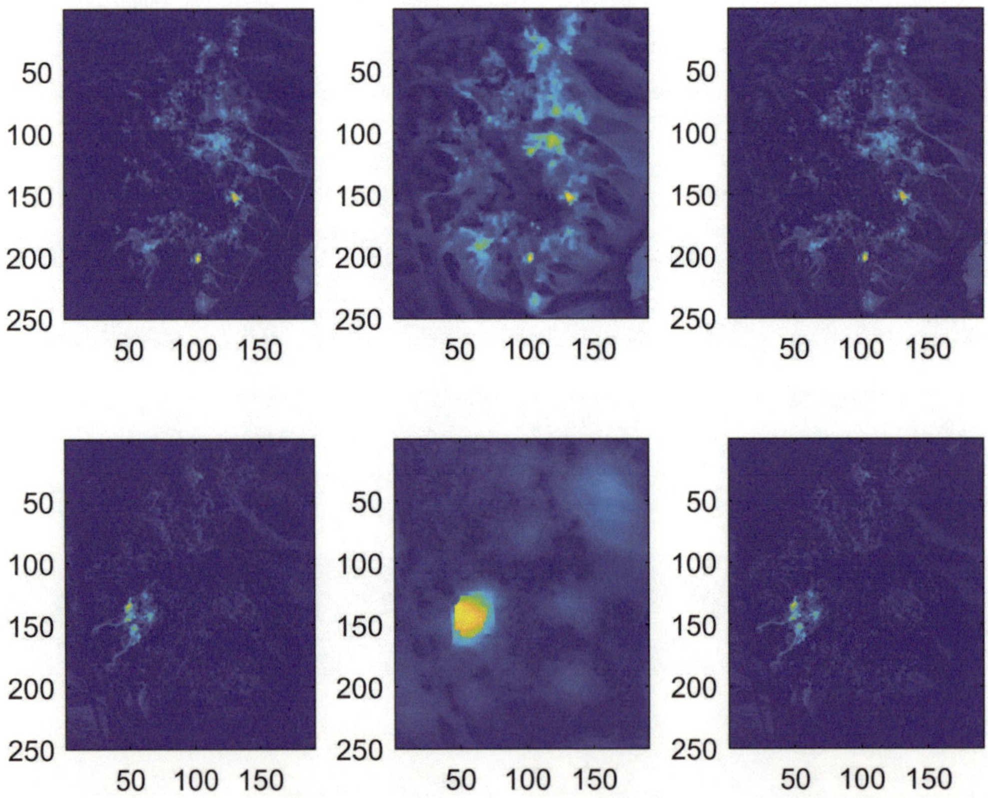

Figure 5. From top to bottom: The abundance maps of first and fifth. From left to right: FCLS, SUnSAL-TV, and the proposed algorithm with the threshold-compared graph.

where all pairs of spectral distance are compared with a user-defined threshold. Meanwhile, the spectral-spatial graph is constructed by considering four neighbors of spatial location and *k*-nearest neighbors of spectral distance.

From this table, we can see that the performance of the proposed algorithm with a threshold-compared graph is better than the others. Although the second graph in **Table 1** combines the spectral and spatial information, using spatial relation is not always a good way to connect pixels because it is possible that two adjacent pixels may have significantly different spectral features. **Figure 4** shows the true abundance maps and the abundances estimated by the proposed algorithm associated with the three constructed graphs. We observe that the second row of the square regions is better conserved with the proposed algorithm using the threshold-compared graph.

5.2. Experiments with AVIRIS data

We also tested algorithms with a real hyperspectral image. The image is captured on the Cuprite mining district by AVIRIS. A sub-image of 250 × 191 pixels was chosen, and it contains 188 spectral bands. The number of endmembers was estimated and set to 12 [26]. VCA algorithm was then used to extract the endmembers. Here, we compare the FCLS [9], SUnSAL-TV, and the proposed algorithm with a threshold-compared graph. **Figure 5** shows the first and fifth abundance maps of three algorithms respectively. We can see that the proposed algorithm highlights localized targets without oversmoothing the image like in SUnSAL-TV and with less impurity than in FCLS.

6. Conclusion

In this chapter, we propose to use graph as a mathematical tool to relate pixels in hyperspectral data. We the present a variety of methods of constructing a graph according to spatial information and spectral information embedded in an image. A sparse unmixing problem is then formulated with the graph regularization to enhance the estimation performance. An ADMM-based algorithm is then presented to solve the formulated problem. In the experiments, we compare the unmixing performance of the presented unmixing algorithm with different graphs, using a synthetic hyperspectral data and a real data. Future works include evaluating the unmixing performance with weighted graphs.

Acknowledgements

This work was supported in part by National Natural Science Foundation of China under grant 61671382 and in part by Natural Science Foundation of Shenzhen under grant JCYJ2017030155315873.

Author details

Zeng Li[1,2], Jie Chen[1,2]* and Susanto Rahardja[1]

*Address all correspondence to: dr.jie.chen@ieee.org

1 Centre of Intelligent Acoustics and Immersive Communications, School of Marine Science and Technology, Northwestern Polytechnical University, China

2 Research and Development Institute of Northwestern Polytechnical University in Shenzhen, China

References

[1] Keshava N, Mustard JF. Spectral unmixing. IEEE Signal Processing Maganize. 2002;**19**(1): 44-57

[2] Dobigeon N, Tourneret J-Y, Richard C, Bermudez JCM, McLaughlin S, Hero AO. Nonlinear unmixing of hyperspectral images: Models and algorithms. IEEE Signal Processing Magazine. 2014;**31**(1):82-94

[3] Singer RB, McCord TB. Mars-large scale mixing of bright and dark surface materials and implications for analysis of spectral reflectance. In: Lunar and Planetary Science Conference Proceedings; 1979. p. 1835-1848

[4] Hapke B. Bidirectional reflectance spectroscopy: 1. Theory. Journal of Geophysical Research: Solid Earth. 1981;**86**:3039-3054

[5] Halimi A, Altmann Y, Dobigeon N, Tourneret J-Y. Nonlinear unmixing of hyperspectral images using a generalized bilinear model. IEEE Transactions on Geoscience and Remote Sensing. 2011;**49**(11):4153-4162

[6] Meganem I, Deliot P, Briottet X, Deville Y, Hosseini S. Linear–quadratic mixing model for reflectances in urban environments. IEEE Transactions on Geoscience and Remote Sensing. 2014;**52**(1):544-558

[7] Altmann Y, Dobigeon N, Tourneret J-Y. Nonlinearity detection in hyperspectral images using a polynomial post-nonlinear mixing model. IEEE Transactions on Image Processing. 2013;**22**(4):1267-1276

[8] Iordache M-D, Bioucas-Dias JM, Plaza A. Total variation spatial regularization for sparse hyperspectral unmixing. IEEE Transactions on Geoscience and Remote Sensing. 2012; **50**(11):4484-4502

[9] Heinz DC, Chang C-I. Fully constrained least squares linear spectral mixture analysis method for material quantification in hyperspectral imagery. IEEE transactions on geoscience and remote sensing. 2001;**39**(3):529-545

[10] Dobigeon N, Tourneret J-Y, Chang C-I. Semi-supervised linear spectral unmixing using a hierarchical Bayesian model for hyperspectral imagery. IEEE Transactions on Signal Processing. 2008;**56**(7):2684-2695

[11] Chang C-I, Wu C-C, Liu W, Ouyang Y-C. A new growing method for simplex-based endmember extraction algorithm. IEEE Transactions on Geoscience and Remote Sensing. 2006;**44**(10):2804-2819

[12] Ammanouil R, Ferrari A, Richard C. A graph laplacian regularization for hyperspectral data unmixing. In: Acoustics, Speech and Signal Processing (ICASSP), 2015 IEEE International Conference; 2015. p. 1637-1641

[13] Castrodad A, Xing Z, Greer JB, Bosch E, Carin L, Sapiro G. Learning discriminative sparse representations for modeling, source separation, and mapping of hyperspectral imagery. IEEE Transactions on Geoscience and Remote Sensing. 2011;**49**(11):4263-4281

[14] Qu Q, Nasrabadi NM, Tran TD. Abundance estimation for bilinear mixture models via joint sparse and low-rank representation. IEEE Transactions on Geoscience and Remote Sensing. 2014;**52**(7):4404-4423

[15] Pirzada S. Applications of graph theory. Journal of the Korean Society for Industrial and. Applied Mathematics. 2007;**7**(1)

[16] Bachmann CM, Ainsworth TL, Fusina RA. Exploiting manifold geometry in hyperspectral imagery. IEEE transactions on Geoscience and Remote Sensing. 2005;**43**(3):441-454

[17] Basener B, Ientilucci EJ, Messinger DW. Anomaly detection using topology. Proceedings of SPIE. 2007

[18] Camps-Valls G, Bandos TV, Zhou D. Semi-supervised graph-based hyperspectral image classification. IEEE Transactions on Geoscience and Remote Sensing. 2007;**45**(10):3044-3054

[19] Stevens JR, Resmini RG, Messinger DW. Spectral-density-based graph construction techniques for Hyperspectral image analysis. IEEE Transactions on Geoscience and Remote Sensing. 2017;**55**(10):5966-5983

[20] Zelnik-Manor L, Perona P. Self-tuning spectral clustering. In: Advances in Neural Information Processing Systems. 2005. pp. 1601-1608

[21] Shi J, Malik J. Normalized cuts and image segmentation. IEEE Transactions on pattern analysis and machine intelligence. 2000;**22**(8):888-905

[22] Gillis DB, Bowles JH. Hyperspectral image segmentation using spatial-spectral graphs. Proc. SPIE. 2012

[23] Bioucas-Dias JM. Figueiredo MAT. Alternating direction algorithms for constrained sparse regression: Application to hyperspectral unmixing. In: Hyperspectral Image and Signal Processing: Evolution in Remote Sensing (WHISPERS), 2010 2nd Workshop; 2010. p. 1-4

[24] Ammanouil R, Ferrari A, Richard C. Hyperspectral data unmixing with graph-based regularization. Proc. IEEE GRSS Workshop Hyperspectral Image Signal. 2015:1-4

[25] Boyd S, Parikh N, Chu E, Peleato B, Eckstein J. Distributed optimization and statistical learning via the alternating direction method of multipliers. Foundations and Trends in Machine Learning. 2011;**3**(1):1-122

[26] Chen J, Richard C, Honeine P. Nonlinear unmixing of hyperspectral data based on a linear-mixture/nonlinear-fluctuation model. IEEE Transactions on Signal Processing. 2013;**61**(2):480-492